Praise for *Borrowing Brilliance*:

"An entertaining, easy-to-read romp through the history of innovation, from Gutenberg to the Google guys, plus a method that appears to actually work." —*BusinessWeek*

"Murray's prescriptions are lucid and helpful, and this book should garner prime shelf space." —*Publishers Weekly*

"Every once in a while I really flip over a new book I get. My latest fave rave is *Borrowing Brilliance*."
—Jack Covert, founder and president, 800-CEO-READ

"David is a practical visionary. He cobbles together two separate worlds: the mysterious, creative world of the artist that generates new forms, and the predefined, cost-effective world of repeatable business success.... In some places it is hard and tight, clearly spelled out. In others, it is more curious serendipity, generated through inspiration and artistry.... I think he has done a great job of coming clean with innovation."
—Seth Kahan, Fast Company

"A wonderfully enjoyable tour of the creative process. Murray's lively description of how he used these tools to create his own products is a major plus. People will be stealing from this work for years to come!"
—Roger von Oech, author of *A Whack on the Side of the Head* and *The Creative Whack Pack*

"Here it is, how to be brilliant and innovative. New ideas are always amalgamations of old ones: borrowed. Murray makes it seem simple, but don't be deceived. Everything he says is true."
—Donald A. Norman, author of *Emotional Design*

"Everything a business book should be. A great concept brilliantly expressed in an interesting and well-written book."
—Al Ries, coauthor of *War in the Boardroom*

"*Borrowing Brilliance* is truly creative brilliance! It is simple, easy to read ... it's magical.... His book brings to life these steps through building on the ideas of others. You can borrow and use his six steps and explode with fresh new ideas the day after finishing *Borrowing Brilliance*."
—Jack Mitchell, CEO of Mitchells/Richards/Marshs and author of *Hug Your Customers* and *Hug Your People*

David Kord Murray began his career as an aerospace engineer working on the conceptual development team for the International Space Station. He has also been an entrepreneur, inventor, and Fortune 500 executive. He was the head of innovation for the software company Intuit and held similar positions at other Fortune 500 companies. He lives in Tahoe City, California.

BORROWING BRILLIANCE

The Six Steps to Business Innovation by Building on the Ideas of Others

David Kord Murray

GOTHAM BOOKS

GOTHAM BOOKS
Published by Penguin Group (USA) Inc.
375 Hudson Street, New York, New York 10014, U.S.A.

Penguin Group (Canada), 90 Eglinton Avenue East, Suite 700, Toronto, Ontario M4P 2Y3, Canada (a division of Pearson Penguin Canada Inc.) • Penguin Books Ltd, 80 Strand, London WC2R 0RL, England • Penguin Ireland, 25 St Stephen's Green, Dublin 2, Ireland (a division of Penguin Books Ltd) • Penguin Group (Australia), 250 Camberwell Road, Camberwell, Victoria 3124, Australia (a division of Pearson Australia Group Pty Ltd) • Penguin Books India Pvt Ltd, 11 Community Centre, Panchsheel Park, New Delhi—110 017, India • Penguin Group (NZ), 67 Apollo Drive, Rosedale, North Shore 0632, New Zealand (a division of Pearson New Zealand Ltd) • Penguin Books (South Africa) (Pty) Ltd, 24 Sturdee Avenue, Rosebank, Johannesburg 2196, South Africa

Penguin Books Ltd, Registered Offices: 80 Strand, London WC2R 0RL, England

Published by Gotham Books, a member of Penguin Group (USA) Inc.

Previously published as a Gotham Books hardcover edition

First trade paperback printing, October 2010

1   3   5   7   9   10   8   6   4   2

Gotham Books and the skyscraper logo are trademarks of Penguin Group (USA) Inc.

The Library of Congress has catalogued the hardcover edition of this book as follows:
Murray, David Kord.
Borrowing brilliance: the six steps to business innovation by building on the ideas of others / David Kord Murray.
p.   cm.
ISBN 978-1-592-40478-0 (hbk.)   978-1-592-40580-0 (paperback)
1. Creative ability in business. 2. Creative thinking. 3. Diffusion of innovations. I. Title.
HD53 M87   2009
658.4'063   dc22   200901939

Printed in the United States of America
Designed by Spring Hoteling

For

Nancy Kord. Nancy Murray. Mom. And Nanny.

CONTENTS

CONTENTS

BORROWING BRILLIANCE

THE SIX STEPS TO BUSINESS INNOVATION BY BUILDING ON THE IDEAS OF OTHERS

Step One: *Defining*
Define the problem you're trying to solve.

Step Two: *Borrowing*
Borrow ideas from places with a similar problem.

Step Three: *Combining*
Connect and combine these borrowed ideas.

Step Four: *Incubating*
Allow the combinations to incubate into a solution.

Step Five: *Judging*
Identify the strength and weakness of the solution.

Step Six: *Enhancing*
Eliminate the weak points while enhancing the strong ones.

PROLOGUE

A LONG, STRANGE TRIP

I lit out from Reno,
I was trailed by twenty hounds.

—JERRY GARCIA

Fifty million dollars.

That's the amount written at the bottom of the contract; the amount a bank is offering to pay for my start-up company; and the amount that would slip through my hands, never seeing the light of my personal bank account. They say that you learn from your mistakes. Well, if that's true, then I've purchased fifty million dollars' worth of insight. Not everyone can say that they have lost so much, so fast, and so magnificently. Unfortunately, this is a true story. It's the story of the search for a creative idea, about its origins, how to construct it, and how it evolves. So, here's what I bought for my fifty million, what I learned from this search: Ideas are constructed out of other ideas, there are no truly original thoughts, you can't make something out of nothing, you have to make it out of something else. It's the law of cerebral physics. Ideas are born of other ideas, built on and out of the ideas that came before. That's why I say that brilliance is borrowed.

Always is. Always was. And always will be.

Go figure, right?

• • • •

As I surveyed the fifty-million-dollar pot on the table, I struggled to compose myself. I didn't want to show my cards, for this was the most important negotiation of my life. It was enough to make a first-round draft pick squirm and so it was difficult to concentrate, as I tried to bluff, tried to close the deal and win the pot. I felt a bead of perspiration form on my brow and hoped he wouldn't notice.

"That's it?" I asked. My hands were damp with tension and I hid them under the table.

"Screw you, Dave," he said.

He was a young guy for a bank president, in his mid-forties, seemed sincere and like someone I could trust. I countered his fifty million with sixty million because I thought that's what you're supposed to do. He laughed. Fifty million was a lot for this start-up and we both knew it. It was twice the offer GE Capital had put on the table a few days earlier. My company, Preferred Capital Corporation, had no debt, since I'd financed it out of my personal savings and its operational cash flow, so most of the fifty million would go right into my pocket. Not bad for a middle-class kid from Massachusetts.

We signed a letter of intent for the fifty million that day. It would be followed by a month of due diligence, the bank auditing Preferred's income statement and balance sheet, followed by the formal transfer of assets and liabilities and a check with lots of zeros in return. It was the fall of 1999 and I was looking forward to a new year, a new century, a new millennium, and a new beginning. There was no need to fear the audit, since I oversaw the preparation of the books myself. It was all just a formality.

I had founded Preferred Capital four years earlier as a finance company that provided loans and leases to other companies that used them to acquire equipment like computers, copiers, furniture, and so on. Preferred would negotiate a contract with the customer, send an invoice to the equipment vendor, and then sell the contract to a bank or GE Capital once the equipment had been installed. For the first few months I was the only employee: president, marketing manager, lead salesman, financial analyst, and receptionist. I had one desk and two phones. I had no intention of creating one of the fastest-growing companies in the United States, I only wanted to be my own boss, make my own decisions, and implement my own ideas. My primary concern was lifestyle, not income or equity or the double-digit growth of my start-up. I didn't want to be part of the rat race, so I moved to Lake Tahoe and started Preferred Capital on the shores of what I considered the most beautiful place in the world.

A hundred years earlier, Mark Twain had explored the same shores and said it was "the fairest view that the earth afforded." Like Twain, I saw the lake on my first cross-country trip from east to west. I couldn't believe such a place existed, and the moment I first saw it, I wanted to make it my home. Composed of deep greens and blues, it's the perfect combination of forests, mountains, and crystal clear water that reflects the cobalt skies above it. The bluer the sky is, the bluer the lake is, and in the High Sierra blue is very blue. For the next twenty years I told family and friends, "Someday I'm going to live in

Tahoe." Someday. So when I established Preferred, a company with a business model that relied on direct mail and telemarketing, one that could function anywhere, I realized I could kill two birds with one stone. I could be my own boss and, at the same time, live within the blue world of the Tahoe Basin. One stone. Two birds. One day became someday and my dream became reality.

However, the company grew faster and bigger than I'd ever imagined. Starting with only fifty thousand dollars, Preferred Capital exploded to over three hundred employees, half a dozen offices, and over twenty million dollars in revenues. Every quarter Preferred grew by 100 percent; financing the growth out of the profits from the previous quarter. I became an expert in expansion—hiring, training, marketing—and crafting the systems to control it. I was in the middle of a moneymaking storm, cash flowing in, around, and out of my company. I became a multimillionaire. It was exciting and I got lost in the hurricane of excitement. It wasn't about the Tahoe lifestyle anymore; it was about being bigger, brasher, and bolder than the next guy. I bought a five-thousand-square-foot oceanfront home perched on a bluff in San Clemente, California, and another one in Crystal Bay, Nevada, perched above the shores of Lake Tahoe. I outfitted each home with Porsches, Range Rovers, and state-of-the-art electronics and gadgets. Sure, I was winning the rat race, but as Lily Tomlin pointed out, I was still a rat.

Then one day as I sat and listened to one of my employees, a pretty young girl in a short skirt, complain to me about the lack of professionalism of another pretty young girl in an even shorter skirt, my mind began to wander and I wondered what it was all about. I wondered what had happened to the dream. I had turned into a bureaucrat trying to negotiate skirt lengths to keep everyone happy. *That's it*, I thought, *time to get out*. I was an entrepreneur, an idea guy, not a manager, and Preferred had grown to the point that it needed management, not a new idea. Let someone else worry about skirt lengths. It was time to sell it, quit the rat race, and get back to the original plan, to the calmer waters of living among the deep greens and blues. I had had enough. It was the late nineties and the Internet bubble was causing a frenzy of mergers and acquisitions. Companies with little or no revenue were selling for outrageous amounts of money. Preferred Capital, on the other hand, had a proven business model and had generated millions in revenues and profits from self-funded growth. It was a very desirable acquisition, and that's why the bank was offering me fifty million.

After the letter of intent was signed, with a month before the formal closing, I decided to take a trip around the world. I packed my bag, my skis, and surfboards, called my brother John, a few friends, and we headed out of LAX in search of adventure. We began with a week in Japan exploring Tokyo, riding the subway at rush hour and drinking sake and eating sushi in the evenings. Then we headed to India and up to the Himalayas to ski the biggest mountains in the world. From there we spent a few days in Kathmandu at the Yak and Yeti Hotel in the shadows of Mount Everest. Then to the savannahs of southern Nepal, hiking and camping, in search of the black rhino and Siberian tiger. Next, to Thailand for the chaos of Bangkok, and finally to Bali and surfing the outer reefs of the South Pacific from the open deck of a small sailboat. This was a prelude to the life that awaited me, the one I wanted, the original dream, exploring the globe without a care in the world, periodically returning to the blue world of Tahoe to regroup.

After a month I came home. It was time to "close" the deal and get back to the original dream. I was summoned to New York City to sign the final papers, hand them the assets and liabilities, get my check, and walk away from the rat race. Forever.

I marched into the Park Avenue boardroom with a fifty-cent Bic pen in my pocket stolen from my hotel room. I had been practicing my signature with it all morning. I thought this would be the last day I ever wore a suit. Adrenaline was coursing through my veins as it had on the slopes of the Himalayas and on the reefs of the South Pacific. This was going to be the most important day of my career, for it would be the last day of my career. I was going to cross the finish line in first place, winning the race that I had found myself running.

"Hello, David," the president of the bank said. It was the first time he had ever called me by my first name. He introduced me to two lawyers, two lawyers without paperwork, without briefcases, and so without anything for me to sign. Something was wrong. Very wrong.

I can't remember how the conversation began or what words were spoken. The memory is lost forever, the way a person loses any recollection of a dramatic car crash. What I do know is this: The bank had been shut down by the FDIC the day before. The bank's mortgage division had invested in too many undersecured lines of credit. By the end of the year the bank was insolvent and out of business. All of this had nothing to do with me or my company. Except that the bank was killing the deal to purchase Preferred Capital and the letter of intent we had signed months ago was now worthless. And to make matters worse they told me it was going to be difficult to fund the thirty million dollars in contracts that had accumulated on my books in the meantime, deals that we had generated, they had approved, and that were now waiting for funding. In other words, I was back in the rat race, this time at the rear of the pack, for I was now thirty million dollars in debt.

I walked out of the boardroom a changed man. Not with a check, but with a bill that was impossible for me to pay. If the transaction had gone through a few days earlier I would not be writing this story.

I called GE Capital, the company that had offered me a paltry twenty-five million dollars a month before, but in an ironic twist of fate, the company president had resigned to become the head of Intuit's TurboTax division. GE now had an interim president who wasn't interested in any deals that his predecessor had started. So, for the next six months my employees and I valiantly chipped away at the pile of invoices, but it was more than we could handle. Like Pickett's Charge at Gettysburg, it was a gallant and honorable effort but one doomed from the beginning. My cash reserves ran out quickly. If I hadn't been so arrogant I would have closed the doors to Preferred Capital the day the bank pulled its offer off the table and walked away with several million dollars, but that thought never crossed my mind. My overconfidence led me to believe that I could fight, market, and manage my way out of the problem. I was an expert at expansion, after all, so I felt confident that I could keep my head above water. But I was sadly mistaken. It just got worse and worse and my world collapsed in on me. It was like I was trapped in a pool of quicksand—the more I struggled, the deeper I was pulled toward the bottom.

By the end of the year my company was ruined, its lines of credit fully extended, and defaulting on its invoices. Two hundred employees were laid off. Many were friends and I felt like I was breaking promises I had made to them. Those who were left sensed the impending doom. Lawyers, guns, and money couldn't save me. I was already sunk.

Unable to cope, my once casual relationship with the bottle turned more intimate. I began to drink heavily. After work I found myself at the local bar, Jake's on the Lake. I took my usual seat at the end of the bar, in the corner, and near the exit so I could make an easy escape if I needed. A friend had told me that vodka and cranberry juice was the drink of choice for the functioning alcoholic, providing the maximum buzz with the minimum hangover. So I made it my choice.

"The usual, Dave?" the bartender asked.

"That's right, Tim. And keep 'em coming. My doctor told me I need at least four an hour." Tim laughed as he poured a very generous Stoli and Cran. I smiled.

"How're you doing?" Tim asked.

"Pretty good," I lied. He had no way of knowing the pain that churned inside of me.

The next day I quietly disappeared from Lake Tahoe. I sold my interest in Preferred Capital for a few thousand dollars and banished myself from the most beautiful place in the world. I had no idea where I was going or if I would ever return. Within a few months I filed for personal bankruptcy and moved into a rented apartment hidden in Tempe, Arizona. As I drove out of the Tahoe Basin in a November snowstorm I glanced back at the lake. It wasn't reflecting blue anymore. It was reflecting black.

Go figure, right?

INTRODUCTION

HONOR AMONG THIEVES

Traveling back in time twenty-five years, I find myself sitting in a waiting room. I recognize it as the Westborough State Hospital in Massachusetts and I recognize myself as a young man with a full head of hair waiting for my friend Sarkis Kojabashian. I call him "Tuna" because his name sounds like "Starkist," the company that sells canned tuna fish, and because I have no idea what a "Sarkis" is. He works at the hospital and I've come to pick him up so we can drive to Cape Cod for the weekend.

Tuna had told me this was called the Westborough State Colony for the Criminally Insane when it opened a hundred years ago. Then it was renamed the Westborough State Insane Asylum. Next it was called the Westborough State Mental Hospital and now it was more cryptically known as the Westborough State Hospital. It smells of medicinal alcohol, damp linens, and dried piss. It gives me the creeps and I hate to think of my friend spending time in here. Sarkis wants to become a psychiatrist and he's working as an orderly to get experience and college credits. A bad idea, I think.

Across from me sits a frail, thin man, middle-aged and dressed in a green hospital-issued smock, like a doctor or surgeon wears. He isn't acknowledging me and doesn't seem to care I'm in the room. He rocks back and forth, mumbling something. I am certain he isn't a doctor. I struggle to hear what he's saying, but can't. He chants to himself, the same thing, over and over.

Where the hell is Tuna? I think. I want to get out of this place. I listen. Now I can make it out, barely, now I am certain what he's saying. He wants something.

"I gotta get a gun," he mumbles. Oh, that's just great, I think. Tuna leaves me in here with a psychotic killer, a leftover from the colony for the criminally insane. This guy's going to pull out a Smith & Wesson from under his smock or attack me with a homemade shank.

"I gotta get a gun. I gotta get a gun," he repeats, faster, louder and more desperately. "I gotta get a gun."

Just then Tuna bursts into the room. "Hey, Murray! How the hell are ya?" he says as he smothers me in a bear hug.

I push him away, pissed off, and motion toward the would-be assassin. "Get me the hell out of here," I say.

"What's wrong?" he asks.

"What'd ya think?" I say as we walk out to the safety of the corridor.

"Oh, you aren't scared of Billy, now, are you?" he asks.

"That guy's nuts."

"No shit. Where do you think you are?" he replies.

Down the hall, faintly, I hear, "I gotta get a gun. I gotta get a gun."

I say to Sarkis, "He's dangerous. He keeps saying that he's going to get a gun."

Tuna laughs and says, "He isn't saying he's gonna get a 'gun.' He's saying he's gotta get some 'gum.' Something to chew on, Murray, not something to blow your brains out with." In the car on the way to the Cape, Sarkis tells me that Billy was admitted to the hospital two years ago. Sadly, he's been saying he needs "gum" over and over for most of that time. As Sarkis understands it, he lost his life savings in a bad business deal, double crossed by his partner, and now finds himself in a dank room, hidden in a mental hospital in an obscure part of New England. Every day Tuna gives him a pack of gum, Juicy Fruit, Big Red, and even Bubblicious, but every day he just repeats the request, over and over, even as he chews away.

Go figure, right?

• • • •

Twenty-five years later, hidden in Tempe, Arizona, I can't help but think of Billy and wonder if he ever pulled out of it. Did he ever stop repeating himself? Did he ever escape from Westborough?

I don't know the answers to those questions. I never will. I do, however, start to wonder about my own sanity. While I'm not incessantly chanting for chewing gum, repetitive thoughts are echoing in my mind, and even though I'm thousands of miles away, I wonder how close I am to being admitted to Westborough. To joining Billy.

I'm consulting for another leasing company, hired to create new ideas. The only problem is, all the ideas I create are just rehashed ones from my glory days. Nothing new. I'm known by my colleagues as an "idea guy," but now every time I sit down to think of one, I keep coming back to Preferred Capital. My thoughts are repetitive, like Billy's, trapped in the past, in a canyon of thought I can't escape. I need some new ones.

I begin to read. Voraciously. The little money I have is being spent on vodka, cranberry juice, and books. I'm reading more than two a week. Books on innovation and creativity. Business books. Books on psychology and philosophy. Science books. Books on neurology and biology. Anything that can help to get the creative juices flowing again. The books seem to work. The vodka does not.

Over the next couple of years I manage to think my way out of the one-bedroom apartment in Tempe and begin a journey out of bankruptcy and into a completely new occupation. I start a small consulting company called Kord Marketing Group, a reference to my mother's maiden name and my middle name, and begin developing new marketing programs for small, medium, and even large companies.

Within a year, I get the opportunity to consult for one of the most prominent software companies in Silicon Valley. While there I come up with an innovative direct-marketing program that dramatically increases retention rates, boosts revenues by fifty million dollars, and adds similar bottom-line profits to the company. In retrospect, the idea would seem so simple and so obvious that the senior managers would scratch their heads and ask, "Why didn't we think of that before?" The founder of the company, a veteran of the Silicon Valley software wars and one of the few to beat Bill Gates at his own game, would find himself more intrigued to know how I came up with the idea than with the idea itself.

"How'd you think of it?" he asked me.

I explained to him how I'd studied his business problem and then looked at how other companies in other marketplaces had solved a similar problem. Then I had constructed the new direct-marketing program out of the borrowed ideas from these other places. It wasn't hard. Once I had the material, it was obvious which pieces would best combine to solve the problem I had defined.

"Cool," he said. He was so impressed by the simplicity of the idea and how I'd come up with it that he created a new position at the company for me. I became the Head of Innovation, a position I hadn't even known existed at Fortune 500 companies, and I was told to come up with new ideas and to teach others in the company to do the same. It was this assignment that led to the book you now hold in your hands.

At first, I was intimidated by my new position. How do you teach people to innovate? Is it even possible? I started to study innovative thinking. As an engineer by training, I was looking for a practical approach to innovation, but everything I read seemed to be shrouded in a fog of mystery. On the other hand, my personal approach to creative thinking was pretty much hack, I just stole or borrowed ideas from other places. In my new position I'd have to develop a more sophisticated approach—or so I thought.

I found that most people believed that creativity was a gift. It can't be taught, they said, it's innate in your thinking process. Either you had it or you didn't. The more I delved, the thicker the fog around creative thinking became. As a subject, innovation was bizarre. The ones who did teach it used words like *synthesis, lateral thinking, empathy,* and *pattern recognition* to describe it. I didn't want to say so outright, but I had no clue what these experts were talking about. I didn't understand—it was over my head. I learned how to moderate a brainstorming session by suspending the criticism of new ideas but quickly realized this was a complete waste of time. The sessions were fun and intellectually intriguing but nothing practical ever came out of them. The more I learned about innovation the deeper into the fog I ventured.

I studied the work of Teresa Amabile of the Harvard Business School. She is one of the country's foremost experts on business innovation and she said, "All innovation begins with creative ideas." Okay, I said to myself, that makes sense, but how do you define a creative idea? What is it? Over time I came up with this simple explanation: A creative idea is one that's new and useful. A new idea that isn't useful, I reasoned, isn't worth much in the business world. I could design a car with square wheels, it would be new and different, but it wouldn't be of much use. Later I'd come to realize that this definition transcended business, for it also applied to science, entertainment, and even the arts.

I continued down this thinking path and asked myself two separate questions. What makes an idea useful? And what makes an idea new? The first question was easy to answer. Since ideas are the solutions to problems, it's your definition of the problem that makes it useful. Solve an important one and you've got a useful idea. Right? The second question, however, was a little more difficult to answer.

To figure it out, I began to study ideas. I looked at my own ideas, the ideas of my colleagues, and the ideas of others in business, science, and the arts. I read biographies of Bill Gates, Steve Jobs, and the Google guys. I looked for the source and form of their new ideas. Then I studied Charles Darwin, Isaac Newton, Albert Einstein, Thomas Edison, and George Lucas. Again, I looked at their ideas. I wasn't trying to determine their thinking processes, I was just trying to determine the structure of their ideas. What made them new and different? It took a while, and I had to wade through a lot of crap, but when the fog finally cleared I realized that each new idea was constructed out of existing ideas. It didn't matter whether it was my simple direct-marketing idea or Einstein's sophisticated theoretical-physics idea—they were both just combinations of existing stuff. Sure, Einstein's stuff was much more complex, but it was still constructed out of borrowed ideas. He even said, "The secret to creativity is knowing how to hide your sources."

Aha, I said to myself. *Maybe I'm not such a hack. Maybe there really is honor among thieves. Maybe we're all thieves.* With this new insight, things became clearer and clearer. I began to tell people: Ideas—not just some but all of them—are constructed out of other ideas. I felt like the kid in the fairy tale "The Emperor's New Clothes" who states the obvious: that the emperor is naked. I began calling bullshit, stating the obvious about creativity and changing the perception of it from a waiting game to an exploration game. In other words, creative thought is the search for an idea that already exists, not the act of waiting for one to pop into your head.

Brilliance, I began to say, is actually borrowed. I learned that this wasn't just a characteristic of modern intellectual life, but has been so throughout human history. Some of the most creative people who have ever lived, such as Isaac Newton and William Shakespeare, were accused of idea theft and plagiarism. It didn't surprise me. Since ideas are born of other ideas, this creates a fine line between theft and originality. In fact, it was during the inquisition of Isaac Newton, after having been accused of stealing in the creation of calculus, that he successfully defended himself with the confession, "Yes, in order to see farther, I have stood on the shoulders of giants." In other words, Newton pled guilty to the obvious, that he built his ideas out of the ideas of others.

As I thought more about this, I came to understand that ideas, like species, naturally evolve over time. Existing concepts are altered and combined to construct new concepts; the way geometry, trigonometry, and algebra combine to form calculus. Thousands of years ago, I reasoned, a Neanderthal man accidentally dislodged a large rock as he climbed a hill behind his cave. He watched as it magically rolled down the slope and he went "aha." The next day he chiseled the first wheel out of another stone and amazed his neighbors with his new invention that he had borrowed, copied, from his observation the day before. Another industrious Neanderthal copied the rock-wheel, except he made it out of a fallen tree, so it was easier to roll. Then another combined the wooden wheel with a basket and created the first wheelbarrow and used it to haul the carcass of a dead saber-toothed tiger. Later, this was borrowed and combined with a horse and a second wheel and the first chariot was created. Two more wheels were added to the chariot and the first carriage was constructed. Ultimately, the horse was replaced with a steam engine to make the first automobile. And so on . . . each new idea being built out of a combination of the previous ones. The more I studied, the more I realized that borrowing ideas isn't just a thinking technique, it's the core thinking technique. The fog was gone. For me, creativity was now obvious and I wondered why the fog had ever existed.

So I began to teach this methodology at the software company where I worked. Then something interesting happened. After a presentation to the CEO and his executive staff, the chief counsel of the company took me aside. "David," he said, "I loved your presentation and I think you're onto something, but you can't teach this to our employees."

"I don't understand," I said.

"You can't teach our employees to steal ideas from other companies," he said. "It's just too risky from a legal point of view. You have to take that part out of your presentation."

I was in shock. How could I teach borrowing ideas without making the obvious connection that your competitors are, often, your greatest source for innovative materials? It was then that I realized why there was so much fog of misunderstanding in the creative process. No one wanted to admit that they were thieves, that at the core of the creative process was the act of borrowing. In order to create, you had to copy. The plagiarist and the creative genius, ironically, were doing almost exactly the same thing. The chief counsel was telling me to disguise the process. He was telling me to put a layer of fog over it so we couldn't be sued in the future.

It was this experience that showed me, firsthand, why the creative process was so confusing and so shrouded in a hazy mist. The fine line between theft and originality was blurring the creative process. Most had a vested interest, like the chief counsel, in keeping the true nature of creativity a secret. I would learn that this wasn't a conspiracy to hide the process so much as it was a natural outcome of an economic- and legal-based society. You see, it was the monetary value in ideas that created the concept of originality. And it was the concept of originality that laid a layer of fog over the concept of creativity.

Let me explain.

Origins of Originality

According to Richard Posner, a judge for the United States Seventh Circuit Court of Appeals and author of *The Little Book of Plagiarism*, ". . . in Shakespeare's time, unlike ours, creativity was understood to be improvement rather than originality—in other words, creative imitation." He goes on to explain that "the puzzle is not that creative imitation was cherished in Shakespeare's time, as it is today, but that 'originality' in the modern sense, in which the imitative element is minimized or at least effectively disguised, was not." In his book he explains that the concept of originality and plagiarism arose during the Italian Renaissance of the fourteenth century. Before this time, it was unusual for artists, architects, scientists, or writers to sign their work. Innovation and creativity were understood to be collaborative efforts in which one idea was copied from another and evolved through incremental enhancements. The concept of plagiarism didn't exist. Copying and creating were rooted in the same thing. The person who copied had an obligation to improve the copy, that was it.

In fact, the term *renaissance* means "rebirth" in French. While we think of the Renaissance as a moment in history when creative thought exploded, at the same time it was an era in which copying exploded, too, for the rebirth was based upon the rediscovery of the ideas of the ancient Greeks. According to art historian and author

Lisa Pon, "If the Renaissance was a culture devoted to finding new ways and orders, it was also a culture inclined to find the roots of that originality in the past." Once rediscovered, the ideas of the Greeks were imitated, recombined, and used to solve new problems. "The challenge of sixteenth-century imitation," Pon said, "was to copy chosen models closely enough for their influence to be recognized, but to diverge enough so that the resulting work was a new one." This is what I mean by the evolution of an idea and what Judge Posner meant when he said that creativity was understood to be improvement rather than originality.

Pon goes on to explain that in the beginning, artists were paid by patrons like the Medici family of Florence. Men like da Vinci and Michelangelo were given room, board, a stipend, and told to create. The focus was on the artwork and not the artist, and so copies were valued just as much as originals. Copying was understood and expected. At the same time, a free market economy was beginning to develop and some of these artists began to break away from their patrons and sell their artwork independently. As this evolved, "by the second decade of the sixteenth century, patrons were often asking for pieces made by specific artists." It was at this time that artists began to sign their work. This gave rise to the concept of "originality"—meaning a piece of art that was created by a specific artist and not copied by someone else. By the end of the Renaissance, there was great value placed on an original, and an artist's signature became extremely important. Copying and plagiarism were now condemned, laying an initial fog of misunderstanding over the creative process. The more valuable the concept of originality became, the thicker the fog became. Artists and writers no longer wanted to share their work but took to defending themselves against copiers and frauds. A dense fog engulfed the creative process and the fine line between plagiarism and originality turned into a gap and ultimately a gaping hole. Today, the chasm is so broadly separated that creativity and copying appear to be contrary concepts rather than the parallel ones that they truly are.

A similar evolution of originality happened in business a century later. At first, goods and services were alike; there was no differentiation between them. Soap was soap and beer was beer. In the beginning, scarcity drove the market. The great differentiator in products wasn't in the product itself, but in the price of the product. The eighteenth-century economist Adam Smith has no mention of trademarks in his concept of a free market economy. The markets, he said, were driven by supply and demand. Products were commodities, copies of themselves. By the beginning of the nineteenth century, business success was driven by costs. Innovation and differentiation was in the machinery and production process and not in the products or marketing of them.

By the middle of the nineteenth century, factories began churning out product for cheaper and cheaper prices. Things like soap were shipped to the local market in barrels and some factories began to stamp these barrels with the same branding iron that ranchers used to mark their cattle. This is how the term *branding* arose and with it the concept of product differentiation. With the advent of packaged goods, the brand mark was placed on the packages, and product originality began to arise. Customers started to prefer Palmolive soap over Ivory soap or vice versa. The products themselves became innovative and the brands, like the artists of the Renaissance, became a valuable asset. Companies that were able to differentiate themselves with creative brands, like Proctor & Gamble, began to win the early brand wars and copying them became unacceptable and illegal.

Once the concept of originality took hold, it was followed by intellectual property concepts like copyrights, trademarks, and patents, which were designed to protect the originator of creative ideas. These concepts shrouded the creative process in a fog of misunderstanding. Today, this misunderstanding results in a creative paradox. We are taught to value creativity and to disdain copying or plagiarism, but copying is the source of creativity. And so we're forced to conceal or disguise the source of our ideas for fear of social or legal retribution. No one wants to admit how they formed their ideas for fear of being labeled a plagiarist or idea thief. The cover-up isn't always intentional, often it's done in the shadows of your subconscious mind. You're unaware of the origins of your own idea, for it magically appears to you in an "aha" moment. But as Einstein said, the secret to creativity is to hide your sources, for he knew the true source of ideas is other ideas—that ideas give birth to one another. That they build on each other. And now you know it too.

In the past, this secrecy and misunderstanding were tolerated because few people made a living off of being creative. Innovation was for a select few like artists, advertisers, entertainers, and entrepreneurs. For most of these people, the creative process took place in the subconscious mind and so it was assumed that creativity was a gift, something you either had or you didn't, it wasn't something that could be taught or manifested consciously. But today, the world is changing. There's a wave of innovation that's just beginning to crest, and before long innovation and creativity will become the responsibility of all of us. I know, because I've been surfing the initial swell.

Let me explain.

Surfing the Innovation Wave

In the book *A Whole New Mind,* author Daniel Pink explains economic evolution using a screenplay metaphor. Economic steps are like the acts in a movie, and the members of society are like the actors in this story. The first act, as he calls it, was the *Agricultural Age,* and the central player was the farmer. To survive in this age one needed a strong back, for work was defined by the hard labor of planting and harvesting the field. The second act began in the nineteenth century and is called the *Industrial Age.* In it the primary actor was the factory worker. To survive in this age the worker tended to the machines and work was defined by long hours and tedious, repetitive tasks. The third act is the *Information Age,* which began in the twentieth century and was dominated by knowledge workers. Most of us are children of the information age for, according to Pink, we're at the tail end of this evolutionary step. To survive in this age the worker gathered and disseminated information, and work was defined by the management of facts and figures. But at the dawn of the twenty-first century, information has become a commodity, and so we're at the dawn of the next step, a step he calls the *Conceptual Age.* The primary actor will be what he calls the "creative" worker. The nature of work will change from the management of existing information to being the creator of new information. The creative worker, in order to survive, will have to know how to ride the innovation wave that's just beginning to crest. You'll need to become the creator of ideas and not just the consumer or manager of them.

For me, the Conceptual Age is already here. In my role as the *head of innovation* for a prominent software company, and later as the *vice president of innovation* for a Fortune 500 services company, the nature of my work is to create new ideas and not just manage or consume existing ones. I suspect that you, even without an "innovation" title, are feeling the pressure to create and innovate just as I do. Product life cycles that were once measured in decades are now being measured in years, even months. Careers once spent in the maternal arms of a single mother corporation are now spent jumping from one company to the next. The need for innovation and creativity becomes more and more important as these product and career life cycles become shorter and shorter. Businesses must be reinvented at a feverish pace to keep up with the market, just as businesspeople have to reinvent themselves to maintain a successful career. Innovation and creativity were once the responsibility of the entrepreneur, the marketing department, or the advertising agency—now they're the responsibility of every employee. Innovation can no longer be outsourced but has to become part of the DNA of every organization. A survey of top U.S. CEOs in *Fortune* magazine listed "innovation" as the primary organizational priority. Or as Tom Peters recently said, "consensus is emerging that innovation must become most every firm's 'Job One.'"

In 1921, as the Information Age was dawning, Claude Hopkins wrote a book called *Scientific Advertising* that became an instant bestseller and the bible for an emerging business discipline called marketing. Up to that point, companies were segmented into sales, finance, and operations, there was no such thing as a "marketing department," and everyone "did marketing." Brands and trademarks were in their infancy and the knowledge workers were just beginning to understand and manage them.

Today, as the new economic age emerges, a new business discipline is emerging with it to meet its unique demands. Instead of "marketing," "innovation" is the new business department. Recently, I spoke at an Innovation Conference in San Diego alongside the *director of global innovation* at Best Buy, the *vice president of innovation* at Whirlpool, the *vice president of innovation* at Raytheon, and a dozen other executives with similar titles and in positions that didn't exist a few years ago. This new group of professional colleagues is testament to the emergence of the Conceptual Age and the importance of innovation in business. It's the result of the evolution of economics, business, and society in general.

Daniel Pink says, "In short, we've progressed from a society of farmers to a society of factory workers to a society of knowledge workers. And now we're progressing yet again—to a society of creators and empathizers, of pattern recognizers and meaning makers." In other words, the innovation wave is coming, and in order to surf it you've got to understand how to construct a creative idea. That's what this book is about, to teach you how to ride that wave and not drown in its wake.

Borrowing Brilliance in the Conceptual Age

The goal of this book is to take the creative process out of the shadows of the subconscious mind and bring it into the conscious world. It's to dispel the misconceptions about creativity, lift the fog off its true nature, and reveal the fact that brilliance is borrowed. In order to create, first you have to copy. Once understood, you can still use your subconscious as a partner in the process, but you'll learn how to take control over it and not sit there waiting for that elusive idea to pop into your mind. Instead, I'll teach you how to go out and find the material for ideas and then how to take this stuff and reconfigure it into a new solution. It's not magical, my friends, it only appears that way. I'm here to tell you that the emperor has no clothes on.

   Borrowing Brilliance is a six-step process, and so this book is organized into six chapters. I think of the first three steps in terms of a construction metaphor. An idea is like a house or a building. Your business problem is the foundation of that house. In other words, you build your idea on a foundation of well-defined problems. Once defined, you borrow ideas from places with a comparable problem. You start close to home by borrowing from your competitors, then you venture farther by borrowing from other industries, and finally you travel outside of business and look for ideas with that problem in the scientific, entertainment, or artistic worlds. Then, you take these borrowed ideas and start combining them to form the overall structure of your house, to form the structure of your new solution. I'll teach you how to use metaphors and analogies to create this structure and so create the overall form of your new idea. I refer to the first three steps as *The Origin of a Creative Idea:*

> Step One: *Defining*
>   Define the problem you're trying to solve.
> Step Two: *Borrowing*
>   Borrow ideas from places with a similar problem.
> Step Three: *Combining*
>   Connect and combine these borrowed ideas.

However, the construction metaphor only extends so far. Creating a new idea requires a process of trial and error, something an engineer or architect would never suggest doing in the construction of a house. So, I think of the next three steps using an evolutionary metaphor. An idea forms over time the way an organic species forms. An idea is a living thing, a descendent of the thing it derived from, the way a rock evolved into the wheel, the wheel into a chariot, and the chariot into the automobile. Ideas give birth to one another. Using this metaphor, your subconscious mind becomes the womb in which new ideas are created. You'll learn how to give birth to them by teaching your subconscious to define, borrow, and combine and so you'll feed it with problems, borrowed ideas, and metaphorical combinations. Then you'll incubate your idea and let your subconscious form a more coherent solution. I'll teach you to use your judgment of this new solution as the mechanism by which to drive the evolution of the idea, in the same way that the fight for survival drives the evolution of organic species. Then you'll separate your judgment into positive and negative, thus revealing the strengths and weaknesses of your new solution. You'll use judgment to improve the idea by eliminating its weaknesses and enhancing its strengths. In other words, you'll create the way the Renaissance masters did, through the incremental improvement of existing ideas. Over time, though, your new idea will grow and evolve, and eventually when you present it to the world it will appear to be completely new and original and the incremental steps will merely be fossils in the process. I call these steps *The Evolution of a Creative Idea:*

Step Four: *Incubating*
   Allow the combinations to incubate into a solution.
Step Five: *Judging*
   Identify the strength and weakness of the solution.
Step Six: *Enhancing*
   Eliminate the weak points while enhancing the strong ones.

The sixth step isn't really a step at all, it's a return to the previous five steps: defining; borrowing; combining; incubating; and judging; all in an attempt to advance your idea through elimination and enhancement. While the first five steps are linear and build off each other, the sixth step is more of a haphazard one. It's more organic, a self-organizing process, one in which the process creates itself and is unique to each project. After passing judgment, you'll return to the problem, reconsider it, perhaps redefine it or decide to solve a completely different one. Your positive/negative judgments will develop your creative intuition and give you greater insight into what to borrow and from where. You'll replace ill-fitting components with new ones that work better. This will help you to restructure your idea and thus make new combinations that work better to solve your problem. You'll simulate the mind of a genius by using left-brain thinking to take your idea apart, reconfigure it, and then use right-brained thinking to put it back together. In between these steps, you'll reincubate, returning to the well of subconscious thought as the process evolves. The order in which you do these things will depend upon your unique situation.

Since I'm not a college professor or academic researcher, this book will not read like a textbook. Instead, I'll use stories to explain my thesis. I'll show you how the Google guys defined their problems in a manner that led to their innovative ideas. How Bill Gates borrowed the ideas of others and created the most powerful company in the world and became known as one of the pirates of Silicon Valley. Then I'll show how Charles Darwin did the same thing but why he isn't called the pirate of Edinburgh Valley. I'll explain how to use metaphors to make combinations, to fuse things together, and create the overall structure of your idea by showing you how George Lucas did this very thing to create his lucrative movie franchise and once you understand, I'll show you how to apply this technique in a practical way in your business. Then I'll tell you the story of Steve Jobs and how he uses his contrasting personality traits to pass judgment on ideas and in the process developed a highly sensitive form of creative intuition. Finally, I'll lead you on this road of discovery by telling you my own story. How I left a one-bedroom apartment in Tempe, Arizona, broken, busted, bankrupt, and with little hope of ever returning to my home in Lake Tahoe. How I discovered the ideas in this book and how I used them to develop my own ideas, to re-create my career, and ultimately to re-create myself. When you're done, you'll agree that brilliance is actually borrowed, easily within your reach, for, really, it's knowing where to borrow the materials from and how to put them together that determines your creative ability. Sadly, I'll never be Steve Jobs, and neither will you, but I can simulate the way he thinks even if it isn't inherent in me. And you can too.

With that said, let's begin the journey.

The Long, Strange Trip Begins

Of course, I don't understand all of this as I sit in my one-bedroom apartment in an obscure part of Arizona nursing a Stoli and Cran and thinking about Billy. I'm praying to God that I won't end up like him, at the same time realizing that I already have because the thought is being repeated over and over in my mind.

I don't know where to begin. How does a forty-three-year-old man re-create his career from scratch? Out of broken dreams? What's the starting point? Surely there must be an answer to that question.

I wonder.

PART I

THE ORIGIN OF A CREATIVE IDEA

CHAPTER ONE

THE FIRST STEP—DEFINING

THE PROBLEM AS THE FOUNDATION
OF THE CREATIVE IDEA

Traveling back in time thirty years, I see myself sitting beside a lake and looking out toward an island. Beside me sits April, my faithful dog and constant companion. She's half border collie and half coonhound; half black and half white; and so a full-bred mutt. I recognize this place as North Conway, New Hampshire. In the distance is Mount Washington. Somewhere in the hills behind the lake is a small cabin, my mom, dad, brothers, and sisters. We're all much younger and better-looking. Dad is still alive, like April.

"What do you think, Abe?" I ask. I call her Abe because she's old and wise. I want to know if she wants to explore the island.

She nods yes. Abe loves adventure.

However, there's a problem: too much water between us and the island. It would be dangerous for us to swim so far because neither of us is a very good swimmer. We'll have to build a raft if we want to be the first man and dog to set foot on the distant shores. And so it's decided, we'll Huck Finn ourselves across the lake and make this the last adventure of the summer.

Back at the cabin I gather my tools as Abe sits and watches. I pack a hammer, rusty saw, and an old ax. All of this goes into my dad's wheelbarrow and we head back toward the lake. Along the way we gather the materials we'll need to solve our problem. From a construction site we borrow some two-by-fours, a small sheet of plywood, a roll of pink insulation, some rope, and a handful of penny nails.

I use the ax to fell a dead tree, laying the trunk down to act as the keel to my craft. Once it's laid, I then carefully place other logs around it to form the hull. I fasten the logs together using the rope. The two-by-fours go on top of the logs, the insulation on top of the two-by-fours, and the plywood on top of it all to form the decking. Abe watches, puzzled, as I construct the solution.

Several hours later I drag the raft into the water, christen her, and set out for a shakedown cruise. She floats, but barely. A boat stays afloat because its weight is equal to that of the water it displaces. Add more weight and the boat sinks a little bit to compensate. Unfortunately, my weight is sinking it so much that the water's lapping up on the decking. Abe will to have to wait for me on the mainland. I'm going to have to explore the island on my own. "Sorry, pal," I say.

As I make my way out into the water I look back and see Abe pacing back and forth, watching, confused, and yapping like an injured coyote. She doesn't understand. I put my head down and paddle, heading toward the island. It's slow going—the raft is about as hydrodynamic as it is seaworthy. It moves through the water like the sun moves through the sky, with no sense of motion. After a while I look back. The mainland is farther away than the island, Abe is gone; I'm making progress, while the sun has crept closer to the horizon.

Like Columbus landing on the shores of San Salvador, when I reach the island I jump from my vessel and claim the territory for myself. I've made it safely across the water, solving the problem I set out to solve. As I pull my craft up on the beach I hear something behind me, a rustling in the woods. *I'm not alone.* Something's coming through the underbrush, intent on attack. I turn around, take a defensive position, and prepare myself for the assault. Fear rises in my bowels and is ready to burst before it registers in my mind.

It isn't an attacker. It isn't a rabid beast. In fact, it isn't a foe at all—it's my faithful dog, Abe. She's rushing toward me, her tongue outstretched and her tail wagging wildly. And she's as dry as a camel at the rear of a caravan. I'm baffled, confused. How the hell did she make it out here?

However, it doesn't take me long to figure it out. You see, my island was not an island at all. It was a dumbbell-shaped peninsula that jutted far out into the lake, connected to the mainland by a thin land bridge not visible from my side of the pond. It wasn't a real island, it was a perceived island. I had constructed an elaborate solution to my problem, a complex raft that took all day to build, when all I needed to do was walk along the shoreline like Abe had done.

Go figure, right?

• • • •

Thirty years later, hidden in Arizona, I'm faced with a new island to explore, a new set of problems to solve. My mind is echoing with repetitive thoughts as I'm struggling to come up with some new ideas, some for my client and some for myself—anything will do. I ask myself: Where do I begin?

The story of Abe and me in North Conway pops into my mind. At first I dismiss it as a random musing and then I realize it's a clue, that it's the answer to my question. I realize I need to begin the search for new ideas with a problem. Over time, I realize every good idea is really a solution to a well-defined problem. More importantly, I recognize that how I define a problem will determine how I solve it—define it as an island and you'll construct a raft, define it as a peninsula and you'll go for a walk. Both ideas solve the problem of getting to the other side, for sure, but who needs an elaborate watercraft when a good pair of hiking shoes will do?

On a yellow legal pad I write: "A problem is the foundation of a creative idea." In other words, a creative idea is built upon the problem one is trying to solve. It's the starting point. This leads me on a new path. I begin studying problem solving, problem identification, and the crafting of problem statements. In a biography of Charles Darwin I read his reflection on the subject: "Looking back, I think it was more difficult to see what the problems were than to solve them." In a biography of Albert Einstein I read his thoughts: "The mere formulation of a problem is far more essential than its solution." To these quotes I add some stories and problem-defining techniques, and write them on my yellow legal pad. After a few months, the legal pad is filled with business ideas, thoughts about thinking, and more insights into the creative process.

Over the next few years the significance of a problem and its role in the creative process become clearer to me. I see the creative process metaphorically as a construction process. I recognize that a problem is the foundation upon which a creative idea is built. Build upon a foundation of sand and your idea will collapse. Build upon solid ground and your idea is much more likely to be realized. Solid ground is achieved by taking the time to study your problem. To do this, I realize, I have to understand the problem with problems.

The First Problem with Problems

The first problem is simple. If you're like me, then you don't take the time to study and construct a foundation. Instead, you charge ahead and construct an elaborate raft of an idea before you properly construct the foundation. Slow down. First, make sure you're building on solid ground.

On my bookshelf sits a popular book on business innovation, one of my favorites. It has more than two hundred and fifty pages of good problem-solving techniques but it has less than two pages on defining a problem. It understands that an idea is a solution to a problem, but spends no time in describing how to find one, understand one, or choose the right one to build upon.

Although you're wired to solve problems, you're not wired to accurately define them. In the fight for survival, which determined your inherent characteristics, the ability to make a quick decision was more important than the ability to make an accurate one. You're wired for speed and not precision. Imagine your ancient ancestor observing the rustling of the grass approaching him on the prehistoric savannah. This was either a saber-toothed tiger or the wind blowing the tall grass. The ancestor who made a quick decision to run was the one who survived, passing this trait to you; the one who stayed to determine the source of the rustling grass was more apt to be eaten by the tiger. His genes, and aptitude for problem analysis, were taken out of the gene pool long before modern times. Speed of thought is in your genetic makeup. It served your ancestors well with life threatening problems, but now causes you to misdiagnose the not-so-life-threatening ones you now face.

Your formal education only reinforced these bad habits. In school you were taught how to solve problems, not how to identify and define them. Your grades were based on the right answers to someone else's questions and, in most cases, how quickly you could answer them. Standardized testing, like the SAT, was scored on the greatest number of correct answers. You never got to formulate your own questions. You learned to solve problems quickly just like your ancestor on the prehistoric savannah. If you wanted to get across the lake you started to build a raft as quickly as possible instead of analyzing the problem and properly defining it.

The clock is always ticking in the background, it seems. The first problem with problems is that it says—Solve it now!—while you've got time. The second problem with problems is a result of this unnecessary constraint. Today I've learned to slow down. I study the problem before I begin constructing an idea.

The Second Problem with Problems

The second problem is a little more complex. You focus on solving a specific problem and forget how that problem fits into a bigger picture. Every problem you choose to solve is part of a myriad of interrelated problems. The second error is in perceiving your problem in isolation and so not understanding its scope. *Let me explain.*

Since you're rushed, you don't take the time to solve the right problem. You focus on the problem at hand instead of understanding the various problems that surround it. Every solution creates more lower-level problems. So every problem is the result of a solution, above which sits a higher problem. It's the nature of problem solving. As playwright George Bernard Shaw said, "Science . . . never solves a problem without creating ten more."

For example, in direct marketing, a high-level problem is the response rate. To solve this problem you construct a mail program comprised of four components: 1) a list, 2) an offer, 3) a package, and 4) the time it is mailed. Each component, in turn, has its own set of problems and solutions. The mail package is comprised of: a) an envelope, b) a letter, c) a brochure, and d) a response form. The "color of the envelope" is a low-level problem, while the overall "response rate" is a high-level one. And yet, they're intimately related; the response rate may or may not depend on the color of the envelope. At the same time, you could completely eliminate the envelope problem by mailing a postcard instead of a letter, solving the higher-level problem in a different way that eliminates the lower-level one. Low-level problems are dependent on higher-level ones and their solutions.

According to Morgan Jones, a former CIA analyst and author of *The Thinker's Toolkit*, the most common mistakes people tend to make when defining a problem are mistakes of scope—defining a problem too *narrowly* or too *broadly*. For example, if you're the vice president of direct mail at a Fortune 500 company and you define your problem as "the color of the envelope," this may be too *narrow* a problem and so not yield positive results. If you define your problem as "response rate," this may be too *broad* and simply lead to the current solutions. These mistakes are the result of looking at each problem in isolation. A *narrow* problem is a low-level one while a *broad* problem is a high-level one.

Every subject presents a series of interrelated problems, often creating a highly complex hierarchy of problems and solutions. It's this hierarchy that forms the foundation you build upon. It's this hierarchy that you ultimately construct in the first step of *Borrowing Brilliance*. Understanding it is how you understand the scope of your problem. This is what Darwin was referring to when he said that it was more difficult to see the problems than it was to solve them (note that he used a plural form to describe problem(s) and not a singular form). In fact, Darwin was well aware of the problems with problems. He took the time to stop and properly analyze his problems and to make sure that he understood their entire hierarchy, not to think of one in isolation. We know this because his notebooks survive today (he labeled them A, B, C, D, and so on) and they're filled with pages and pages of questions and problem definitions. High-level ones and low-level ones.

Constructing the Foundation

Every journey begins with a first step, just as every construction project begins with an initial task. Since a problem is the foundation of a creative idea, your first task in constructing that idea is to construct the foundation itself. Constructing a foundation for a building involves choosing a site, laying the footings, and then pouring the foundation upon those footings. Similarly, constructing an idea foundation involves: identifying a problem (choosing a site); determining root cause (laying the footings); and understanding the scope (pouring the foundation). The creative genius is constantly in search of problems, realizing that ideas are solutions, and so has an innate ability to identify and understand them. The genius does it naturally in the subconscious mind. You and I, on the other hand, have to be more deliberate and do it consciously.

In my notebook I record the stories of creative thinkers. It helps me understand the creative construction process from a practical standpoint. It'll help you to better understand too. So, imagine that it's a beautiful January afternoon in 1996 at Stanford University's Palo Alto campus. A small group of teachers, students, and alumni have gathered on the southwest side of the grounds to dedicate a four-story building. It's beige stucco with a red tiled roof that architecturally blends with the other mission-style buildings that dominate the school. Over the building's main entrance an inscription, chiseled in stone, reads: WILLIAM GATES COMPUTER SCIENCE. Bill isn't there but the dean of the engineering school, James Gibbons, is, and says a few prophetic words: "Something will happen here, and there will be some place, some office, some corner, where people will point and say, 'Yeah, that's where they worked on the *blank* in 1996 and 1997. And, you know, it was a big deal.'"

That place, as it turns out, is a room on the third floor, in the western corner of the building. For in Gates 360, two students would lay an idea foundation upon which they would build a business that would dominate their competitors, the Internet, and international commerce for years. Ironically, the problems they would choose to solve would allow them to become the chief rival of the business that Gates himself had founded several decades earlier in another room in another university in another state.

Sergey Brin was born in Moscow in 1973 and emigrated to the United States in 1979. His parents were mathematicians. Larry Page was born in Lansing, Michigan, also in 1973 and also to mathematicians. Larry and Sergey were both brilliant students, Sergey finishing high school in less than three years. When Sergey and Larry moved into their office in Gates 360 they instantly became friends. In fact, they became known around campus as LarryandSergey. Today they are better known as the founders of Google and often referred to as the Google Guys.

In Gates 360 they discussed many problems. Since Larry was from the Detroit area, he was interested in automating transportation systems. According to Sean Anderson, one of the other members of Gates 360, Page "liked to talk about automated automobile systems where you have cars that will roam around, and if you need one, you just hop in and tell it where it needs to go. It is like a taxi, but it is cheaper and packs itself with other such vehicles on the freeway much tighter." Andersen also said that Larry "is passionate about the problem of moving things or people around. He liked solving the problems of society in various ways."

Of course, transportation automation was not the site they chose to construct their ideas upon. I mention it here merely to illustrate that a creative mind is constantly in search of problems, for the creative mind realizes every idea is the solution to a well-defined problem. Choosing a construction site is a matter of trial and error. You've got to look at different sites before making a decision on where to build, on where to construct your ideas. You've got to find solid ground, an important problem to solve and one you've got the skills to solve. As luck would have it, Larry and Sergey independently stumbled upon the same site at roughly the same time. As Sergey put it, "The more you stumble around, the more likely you are to stumble across something valuable."

Identifying a Problem

Innovation begins with the identification of a problem. Innovators are constantly looking for them or asking for them. It's often more difficult to see the problem, as Darwin said, than it is to solve it. So, as you venture out, become more aware of the problems that surround you and your subject. It may seem counter intuitive to look for problems, but it isn't.

As a Ph.D. candidate in computer science, Sergey was assigned a data-mining project working for a Stanford professor. Data mining is the process of sorting through large amounts of data and picking out relevant information. Larry, also working on his Ph.D., was assigned to work with a different professor on a plan called the Digital Library Technologies Project. Its goal was to design and implement the infrastructure for creating, disseminating, sharing, and managing information in a large digital library.

As Sergey and Larry worked on their respective projects, they independently became sensitive to the problems of searching and finding relevant information. Sergey was developing algorithms that businesses could use to extract buying data so they could more efficiently stock retail store shelves. Larry was developing algorithms that researchers could use to find information in a vast database of digitized books. While each was working on a different low-level problem, they quickly realized they were working on a very similar high-level problem.

Late nights in Gates 360 were fueled by cold pizza, warm Red Bull, and the constant banter between Larry and Sergey. Tamara Munzner, the only female among the 360 crowd, had to tune out the relentless chatter and said, "I learned to program with headphones on." During these discussions Larry and Sergey began to scout out the location for their idea foundation. You see, they both spent enormous time researching their respective Ph.D. projects using the Internet to gather data and cite new sources of information. Therein lay the problem, and therein lay the idea foundation: Search engines sucked.

At the time, there were dozens of different engines like Lycos, Magellan, Infoseek, Excite, and AltaVista. Each was designed to solve the problem of finding information on the vast network of Web sites known as the World Wide Web. Two other Stanford doctoral students had identified the same problem a few years earlier and created a relatively simple solution. Rather than relying on technology, Jerry Yang and David Filo solved the problem by establishing a Web site to be used as a portal in which editors selected content and created lists of directories by grouping things under high-level categories (like sports, entertainment, news, finance, and so on). They called this solution Yahoo! It was becoming an Internet darling at the time that Larry and Sergey were eating cold pizza and drinking warm Red Bull, and struggling with the inferior search engines.

At some point the Gates 360 discussions became more and more focused on search engines and the problem of finding relevant material on the Internet. As Larry and Sergey staked out a site for the construction of ideas, Tamara was forced to turn up the volume on her headset to drown them out. The Google guys had identified an important problem and one that they were well qualified to solve.

Their mathematical backgrounds, deep knowledge of computers, and Ph.D. focus on searching and organizing information made them the perfect architects for designing a better search engine.

For Larry and Sergey, identifying problems is a natural part of how they think. It's wired into their minds. For Sergey, this ability was honed early on in his education, for he went to a Montessori school. Instead of taking an endless series of quizzes and tests, he was allowed to identify his own problems and work on them at his own speed (which happened to be very, very fast). Ironically, Larry also attended a Montessori school and so also knew how to define his own problems. And so while it's natural for them to seek out problems, I suspect that it's not as natural for you and me. We've been trained to answer questions, not formulate them. We've been trained to take tests, not construct them. And we've been trained to avoid problems, not to go out and seek them.

Now that you've decided to become more creative, you're going to have to learn how to think more like Larry and Sergey. You're going to have to become the seeker of problems and not just the solver of them. The innovator either asks for them or looks for them, what I refer to as assignment and observation. Both tactics are used to look for the problems in your subject. Each is used to choose a construction site upon which to build your creative ideas. They're used independently or together the way a hammer and chisel work for the craftsman constructing an intricate foundation.

Ask for a problem to solve. Ask your boss. Ask your spouse. Ask your business associates. Or ask yourself. Larry and Sergey were both assigned problems as part of their Ph.D. program. These assignments began their search for a construction site. Problem assignments allow you to scout out a place to build your ideas. Your understanding of the problem will evolve as you identify the high- and low-level problems that surround it and construct the problem hierarchy. Use assignments to place yourself into that hierarchy.

Some of mankind's greatest ideas began as problem assignments. In 1685, Sir Edmond Halley, of comet fame, traveled from London to Cambridge to seek out a young mathematics professor at Trinity College. Could this young man use his new mathematical tools to prove or disprove Kepler's Law of Planetary Motion? Intimidated by the highly cultured and famous Halley, the disheveled young man accepted the assignment and began working on the problem. A few months later, he sent his solution by courier back to Halley at the Royal Society in London. As Halley sat in front of a warm fire, sipping bitter British coffee, he realized, almost immediately, that he was reading something of extraordinary brilliance and intellectual significance, one of the greatest bursts of creative thinking that the world had ever seen. Not only had the young man solved the specific problem that Halley had assigned, but being a creative genius he had used the assignment as an entry point into a far more complex hierarchy of interrelated problems. The shy young man had proven Kepler's Law, but then had expanded his solution to encompass more than just the planets. His solution explained the movement of all objects big and small, those in the sky and those on the ground. With a stroke of brilliance, he had solved an intricate matrix of problems. Important problems. Ones that would change the world. The obscure professor was Isaac Newton and the solution was, under Halley's insistence, expanded, documented, and published in a book called *Principia Mathematica*. Hundreds of years later I would study these same ideas as an engineering student and use Newton's equations to help design and build an earth-orbiting space station as an aerospace engineer.

In business, problem assignments define the relationship between a manager and the managed. For example, the vice president of product development for a manufacturer sets department goals and then works with direct reports to identify problems that must be solved in order to produce products on time and under budget. As a consultant, I ask my clients to be very specific about the problems they want me to solve. We spend hours discussing, analyzing, and defining the hierarchy of problems before we ever sit down to solve them.

Assignment is merely your entry point into the overall matrix of problems, just as Newton used Halley's assignment as a way to enter a far more complex hierarchy of interrelated problems. It's a starting point. The site on which to build your ideas.

Observation, like assignment, is used to seek out problems and so provide an entry point into the problem matrix. Like Newton, the creative genius is an observation expert, adept at gleaning insight into the hierarchy of problems by taking the time to stop and examine the world as it unfolds. While creative observation comes naturally to people like Newton, you and I have to be more deliberate. We have to emulate what he does. So, what makes Isaac Newton realize that an apple falling from a tree, an observation millions have made before him, is a critical problem and part of a more significant matrix of problems?

Put simply, it's this: Observation is the act of studying the production and destruction of patterns. The creative genius recognizes patterns, is always conscious of them in all their glorious forms, and then observes either the making or breaking of them. This is exactly what Isaac Newton did as he sat in his mother's garden in Lincolnshire. As legend has it, first he noticed the apple on a tree was the same shape as the moon in the sky. They made similar patterns (round objects). Then he watched as the apple fell to the ground, noting that it was being pulled toward the earth, falling. He picked it up and held it toward the sky and saw that the moon was moving around the earth, rotating, and not falling the way the apple had. This was a breaking of the pattern that the apple had created in his mind. One was falling and the other was rotating. This observation turned into a problem: Why did the apple fall and the moon rotate? It was this observation, this problem, that led him to construct an equation that explained the behavior of the apple and moon, an equation to calculate gravity and its effects.

Larry and Sergey, like Newton, used the observation tool as they began constructing the foundation for their ideas. As Sergey worked on his thesis he became more and more adept at recognizing patterns. In fact, according to him, "We were doing research in managing large amounts of information, and what's called 'data mining,' which means finding patterns in them." At the same time, as Larry worked on the Internet he observed the output from different search engines and found it hard to detect a pattern in the results. They seemed random, at best. However, he did notice a different pattern in the results that the AltaVista engine returned. AltaVista listed the number of links that each Web site contained, something no other search engine was doing. A link, as you know, is a "hot" piece of text or graphic that takes you to another page or another Web site. It was this observation, a simple break in the pattern of search engine results, that would ultimately lead to the innovative algorithm that would be the cornerstone of the Google empire.

Scott Cook, founder of the software company Intuit, used an observation to create his first product, Quicken. He watched his wife struggle to balance the family's checkbook every month and wondered if there wasn't a better way to solve that problem. Later, when his company became a Fortune 500 powerhouse, he established the "Follow Me Home" program in which an employee would "observe" a customer using an Intuit product in a real-world setting. A number of Scott's people noticed that customers were using Quicken to manage their small businesses, breaking the pattern that the product had been designed to create. This led to an even more important product, QuickBooks, the accounting software, which is now the company's flagship product.

Observation is the basis of the scientific process. As philosopher Karl Popper said in his book *All Life Is Problem Solving*, observation is the primary scientific tool, the beginning of all knowledge. The scientist makes an observation of natural phenomena and then tries to explain the observation through hypothesis. Darwin observed the making of patterns, like common structural elements among a wide variety of different species, like the way a human hand is structured similarly to a bird's wing. Alexander Fleming observed a consistent pattern of fungus on the culture dishes he left in his laboratory. Then he noticed the breaking of a pattern, that one dish had a zone around the fungus where the bacteria didn't seem to grow. First he noticed the making of a pattern, the fungus, then he recognized the breaking of that pattern, the absence of fungus. He set the dish aside and later used it to isolate an extract that destroyed bacteria. That extract became known as penicillin.

Now as you study your subject, you can become more aware of the making and breaking of patterns and so begin to think more like Newton, Darwin, or Fleming. Larry and Sergey do. Use observation to identify problems and begin constructing your idea foundation.

Determining Root Cause

After you've found a place to build your ideas, next you prepare the site for construction. You've identified an important problem, one you're capable of solving. But before you begin building on the problem, you have to take the time to understand it. Study it before you place any weight on top of it. It's like laying the footings for a foundation. The footings for a large skyscraper are concrete and steel pilings that are driven deep into the soil so that the building rests on solid bedrock, far beneath the surface, and not on the less stable topsoil that may buckle under the weight of a large structure. For us, this means understanding the root cause of the problem.

Several years ago my friend David Meyers talked me into climbing Mount McKinley. Alaska natives call it Denali, and it's the biggest mountain in the world when measured from its base to its summit. To reach the top we would have to climb a steep headwall at fifteen thousand feet, a knife-edge rock ridge at seventeen thousand feet, and a knife-edge ice ridge at twenty thousand feet. While not particularly technical by mountaineering standards, it is nonetheless quite challenging for someone who has a fear of heights and is subject to vertigo. To me, walking across the Golden Gate Bridge was terrifying, making me dizzy and disoriented. The final approach to the summit of Denali is via a two-foot-wide arête with a thirteen-thousand-foot drop just inches from the route, certainly not a good place to be disoriented. I had to solve my vertigo problem before attempting McKinley.

So I searched for the root cause. I read books. I talked to rock climbers. And found myself scrambling in high places. Over time, I realized that vertigo is a matter of misplaced focus. Standing beside a cliff, my mind would focus on the empty space just a few feet away, causing me to move toward it (you'll subconsciously move toward whatever you focus on). In response to this subtle movement I would subconsciously push myself away. The end result is an uneasy feeling of instability, of moving toward and then away from the edge. This alternating movement was the root cause of my vertigo.

Once understood, I began to train my mind through repetitive exercises to stay focused on the ground and not the open space. I practiced by suspending a ladder between two rocks in my backyard, walking across it hundreds of times, while consciously focusing on the ladder and not the ground. Months later I found myself walking the final approach to the McKinley summit using the same focus I had developed in my backyard. I made it. Of course, I still have a fear of heights. But since I've solved the vertigo problem I'm able to safely climb in high places. I don't get dizzy and disoriented anymore.

It's no different in business, science, or the arts. You can't build a so-lution on a weak foundation, just as you can't solve a problem without understanding its root cause. I use one very simple tool to do this: I keep asking, "Why?" This tool is used to construct the footings for the intel-lectual foundation of my ideas. I've learned to spend a lot of time under-standing my problems before I jump in and start solving them. I've learned to walk around the lake and study the island before I begin con-structing an elaborate raft to get across to it. So, too, should you.

Curiosity is a common trait among the creative, and consciously using this tool will help you to increase yours. Rajeev Motwani, the thirty-year-old Stanford advisor to Larry Brin on the Digital Library Project, recalled that Larry was a deep thinker and was always trying to determine why things worked. Whether Larry thinks of "why" as a defining tool or not I don't know, but his curiosity is legendary and based, in part, on determining the root cause of his problems.

Like most children, when my daughter, Katie, was younger she'd bombard me with why-based questions. Her curiosity was conta-gious, albeit sometimes annoying. In her defense, though, she was just trying to determine root cause.

"Katie, you need to get ready," I'd say.

"Why?" she'd ask.

"We're going to the store."

"Why?"

"We need to get milk."

"Why?"

"We're out of milk."

"Why?"

"You've been drinking it."

"Why?"

"You love milk. And you're going to want some more this eve-ning when I make you chocolate chip cookies."

"OK, Dad."

And so you can borrow this tool from Katie and use it with your own problems. Just keep asking, "Why?" until you get to the source. I've heard it said that you can always get to the root cause of any problem with just five "whys." I'm not sure if that's true, for as I recall Katie could keep it going for a lot more than just five "whys." *Go figure, right?*

For the Google guys, once they'd identified a problem with search engines they began to ask, "Why?" Sergey asked Larry, "Why are the results so bad? Why when I enter *Stanford* into AltaVista do I get a slew of porn sites and not get the university's Web site?" The problem, they'd come to realize, was that the AltaVista algorithm couldn't determine the importance of different Web sites, it merely returned those sites that had the word *Stanford* contained on a page. A porn site operator could jury-rig the system by repeating the word *Stanford* over and over on his Web page and get a top ranking for his Web site.

Once they understood the root cause of the problem, Larry and Sergey were in a better position to build a solution. Root cause put their understanding on solid ground, bedrock on which to build their ideas. It changed how they thought about the problem. Now they had to find a way to determine the importance of different Web sites. This definition would lead them to the observation that Larry had made earlier in the creative process: the links that AltaVista was recording.

He realized that links were a possible solution to the problem. According to Larry, the more pages that were linked to a specific page, the more important that page. Links were like the bibliography in a book or citations in a research paper. "Citations are important," Page said. "It turns out, people who win the Nobel Prize have citations from ten thousand different papers." A large number of citations "means your work is important, because other people thought it was worth mentioning." And it wasn't just the number of links but also the number of links from "important" pages. For example, Yahoo! was important because it had a huge amount of traffic, and so getting a link from a Yahoo! page was more critical than getting a link from David Murray's home page which saw little or no traffic. It was these factors, and others, that Larry and Sergey began to consider as they started building the foundation and structure to solve their problems.

You'll want to determine the root cause of every problem you identify. Solving your problem is sometimes just a simple matter of eliminating the root cause. In the 1960s, traffic became so snarled on the Los Angeles Freeway system at rush hour that the term *gridlock* was created to define it (since *rush hour* didn't seem appropriate anymore). The root cause was all of the cars attempting to get on the freeway simultaneously. So the city staggered working hours and installed stoplights on the freeway ramps, regulating the cars entering the system. It worked until these ideas were overwhelmed by even more automobiles. (Today, they need a new solution.)

I should have used this tool as I sat on the shores of the small lake in North Conway, New Hampshire. By asking, "Why?" I would have highlighted that I didn't really know if the land across the water was an island or not. This question might have prompted me to explore a little more and so not build my solution on such an ill-conceived foundation.

Now that you've identified a problem (chosen a construction site) and taken the time to understand the root cause (placed the footings for the foundation) the final step is to define the complete set of problems that surround the one you've identified (and so pour the concrete for the foundation).

Understanding the Scope

You can't build a solution on a single isolated problem. You've got to build on the entire matrix of problems: the high-level ones and low-level ones. Building a solution on a single problem is like trying to construct a building on a single piling. It doesn't work very well and in the end your solution will be unstable.

So, you've got to construct that hierarchy—it's the only way to understand the scope of the problem you choose to solve. The scope is the relationship between a single problem and its place in the matrix. There's a number of different ways to do this. Using the identified problem as a starting point, list all of the associated problems relative to this one. If you're working with a team, then hold a series of meetings in which you brainstorm problems (not solutions). Try and think of all the problems connected with the construction site you've chosen to build upon.

When I worked for a software development team we spent several weeks defining, sorting, grouping, and finally arranging all of the problems associated with our product, our marketing of it, our customer, and our own internal issues at the company where we worked. We included technical problems, marketing problems, and even organizational problems.

As a member of the concept development team for NASA's space station we did the same thing. We spent months defining, sorting, grouping, and finally arranging the problems related to our future spacecraft. These problems included constructing and launching the spacecraft, living in it, working in it, as well as the problems associated with our own company and working with a government agency like NASA.

I've found that it's best to begin this process of determining your matrix with a data dump. Just let 'em fly. Don't worry about the significance of the problem. Don't worry about where it fits, or whether it's a high-level or low-level problem. Just identify as many as you can.

Once you've got all of these problems, group them into categories. Grouping puts "like" things together. It helps you to determine if you're missing something. For example, if you've got a "search" problem, like Larry and Sergey, then you'll ask: *Are there any other search problems I've missed?*

Grouping also allows an easy transition to the next step in *Borrowing Brilliance*—gathering materials to solve your identified problem. You want to borrow from places with a similar problem. If you've got a group of "navigation" problems, then ask yourself: *Who else has navigation problems?* You'll answer: sailors, pilots, truck drivers, explorers, and rats caught in a maze. Then you'll look to see how each of these groups solves its problems. Therein lies the key to *Borrowing Brilliance*. You'll learn how to borrow ideas from people, places, and things that have the same kind of problem as your people, places, and things. Who knows, maybe a rat can teach you something. This is why how you define problem will determine how you solve it—for the solution will be found in places with a similar problem.

Once grouped, then you start sorting or rank-ordering your problems from highest to lowest. For Larry and Sergey, the high-level problem was finding information on the Internet. The midlevel problem was creating the algorithm to find the most important pages. And the low-level problem was acquiring the hardware, the computers they'd need to analyze the entire Internet so as to determine the page ranks and the search results. You can't tell someone where to find information unless you've already looked yourself. Of course, there were dozens, perhaps hundreds, of other problems associated with those mentioned. Some were formally documented as they began the process of solving their problems and some were merely floating in the minds of Larry and Sergey.

Use the grouping process no matter how complex or how simple you perceive your problem to be. Creative genius does it naturally, often subconsciously, but you and I have to do it deliberately, consciously. For example, if you're a direct marketing manager, this process begins with a sales goal. Next comes the response rate. Then list, offer, package, and timing problems. And so on down the line until you select the wording on the offer, the color of the package, and the exact date of the mailing.

The grouping process helps define the scope of problems and aids in avoiding the mistake of being too broad or too narrow in your approach. In other words, use the list you come up with to choose the starting point in the problem-solving process. The higher-level solutions will naturally resolve the lower-level ones. But, as you now know, a new solution will create a new set of lower-level problems. That's why problem solving and creativity is a process of trial and error. Whatever solution you construct will, undoubtedly, result in some other problems.

I've developed a simple set of tools for completing this final step. In order to understand the scope I simply take a problem and then look above it and below it for the problems that surround it. Looking up means determining the problem above the one in question by identifying the solution that caused it. Looking down means determining the problems I will create by solving the problem in question.

First ask: *What problem was solved that led to the solution that created the current problem?* This illuminates the higher-level problem and determines the scope of the original problem. If I had stopped to consider the higher-level problem in North Conway (getting to the other shore) before I built my raft, a lower problem, I might have considered other alternatives and it might have occurred to me to walk to the other side. Instead I focused on the subordinate problem of building an elaborate watercraft.

Imagine you work for the Ford Motor Company and your boss gives you a problem to solve—prevent customers from losing the keys to the car. Focusing on this problem in isolation will lead to solutions like bigger keys, a key chain with a flashing light, or keys that respond to a clap with an electronic chirp. However, using the Looking Up technique, you change the scope by asking: *What problem are the keys designed to solve?* The answer is starting the car. So you could solve this higher-level problem by allowing customers to start their car with a combination lock, eliminating the key, and solving the lower problem of losing them. Starting the car is a high-level problem, losing the key is a lower one.

As Larry worked on his Digital Libraries project, he looked up his problem hierarchy and realized he was working on a search problem. Sergey did the same thing. The problem above his data mining problem was also a search problem. So, when they decided to stake out a claim on the Internet it only made sense that they would choose "search" as the site to build their ideas upon.

After you look up, pause for a moment, then look down. Identify a specific problem and ask yourself: *What problems do I create when I solve this one?* Lower-level problems are the result of higher-level solutions. The color of an envelope (low-level) is the result of choosing to send a letter to solve a direct marketing problem (high-level). Since every solution creates new problems, you need to identify them before committing to the implementation of the solution. I've solved many high-level problems only to create a series of lower-level ones that were more detrimental than the original one. This is not unusual in business; you solve a problem for one customer only to create another one for a different customer. Asking this question helps you to perceive the wave of problems you are about to let loose with a certain solution.

For example, I worked with a software development company

that kept adding features to their product to solve specific customer complaints. Over time, the product became more and more difficult to use—customers couldn't find the core features hidden in the long list of supplementals. A lot of electronic products, like my cell phone, have the same disease. It can't make up its mind whether it's a telephone or a computer.

Lower-level problems are less complex and so typically easier to solve than higher-level ones. It's easier to choose the color of an envelope than to try and construct an innovative direct-marketing program from scratch. But the devil is always in the details. While typically easier to solve, they can still make or break the higher-level solution. If I'm selling sophisticated business software in a pink envelope, this may ruin the entire marketing program, even the well-thought-out higher-level solutions.

Don't disregard a problem just because it ends up at the low end of the list. Low-level problems must be solved. I'm certain that Captain FitzRoy of the HMS *Beagle* considered his problems—feeding his men, not running aground, mapping uncharted waters—far more important than those contemplated by the young Charles Darwin, who was just cataloging flora and fauna as his ship anchored off of the Galápagos Islands.

Remember, the creative genius is aware of the full scope of problems and realizes that inconsequential low-level ones are ultimately connected to monumental high-level ones. As Larry looked down he saw another problem below his search algorithm problem. Once developed, he'd have to index the entire World Wide Web by downloading it onto his computer. While this was a lower-level problem, it was a monumental one to say the least. Larry started telling his friends, teachers, and fellow students that he was going to copy the entire Internet onto his PC in just a few days. It became a rallying cry in Gates 360. People thought he had gone insane. He said, "I got this crazy idea I was going to download the entire Web onto my computer. I told my advisor that it would only take a week. After about a year or so, I had some portion of it." Tamara simply turned up the volume on her headphones again to smother the incessant banter.

Choosing the Problem

Now that you've established your foundation you can begin the task of solving these problems. As you isolate a single problem to solve, remember, it's part of an overall system of problems, high ones and low ones, and the most elegant solutions are those that use the least number of components to solve the greatest number of problems in the matrix.

The elegance of Isaac Newton's solution to Edmond Halley's problem was, in part, the fact that he had solved multiple problems with a single equation. He had combined the works of Galileo with the works of Kepler to solve the higher-level problem of gravitational attraction. Galileo had studied how objects behave on the surface of the planet by dropping things from the Tower of Pisa. Kepler had studied how the planets behave by observing their movements in the night sky. Newton realized that these problems were related, part of the same hierarchy of problems, laddering up to a higher problem. So, when he solved the higher problem he subsequently solved the lower ones, therein creating one of the most elegant solutions ever devised.

In 1864 James Maxwell did the same thing. Using the same equations, he developed a theory that solved both: 1) electrical problems; and 2) magnetic problems. One solution, two problems solved. Then Einstein did it again with his relativity theory, using a single solution to explain both 1) electromagnetism; and 2) gravity, solving the problems of both Newton and Maxwell. Einstein actually solved four problems with one theory and so goes down in history, with Newton, as one of the most elegant problem solvers ever. He had a complete grasp of the complex hierarchy of physical problems that surrounded him.

Of course, you and I aren't going to solve such complex problems. We're going to be working on marketing problems or product development problems or career enhancement problems, not trying to describe the universe through a complex mathematical language. However, whatever you study will still present a hierarchy of problems and so you still need to have awareness of these problems. Your creative solutions will become more elegant, often simpler, and more effective. You'll develop solution efficiency.

To take the next step, you're going to have to look at your foundation, your problem matrix, and decide which problem to start solving. Do you start high? Or start low? Starting high is typically a lot more work. It means a big change in direction. As a marketing director this means having to decide whether to use direct marketing or television advertising to solve your sales problem (a high-level one). A lower-level problem would be deciding whether to send a postcard or letter to a prospective client. However, this decision, whether to start high or low, isn't as hard as it may seem.

To do this, think of your problem matrix as a chain of problems. Then ask yourself: Where is the weakest link in the chain? Which problem is going to be the most difficult to solve? And if I can't solve that one, is the entire chain going to be worthless? Remember, a chain is only as strong as its weakest link. Look for those weaknesses.

For the Google guys one of the weak links was the hardware. Since they were working out of a one-room office, and given the small amount of computing power available to them—they had access to only a single computer—it was going to be very difficult to index the entire World Wide Web. From a conceptual level, the hardware was a lower-level problem, but from a practical level it was the weakest link in the chain of problems. That's why Larry made it his rallying cry. So, while they started developing the algorithm to determine Web site importance, they also started downloading the entire Internet by building their own computers out of pieces of computers borrowed from the loading docks at the university. According to Sergey, "We would just borrow a few machines, figuring if they didn't pick it up right away, they didn't need it so badly."

The mistakes in problem solving, according to Morgan Jones, are mistakes in scope. They're mistakes in choosing the wrong problem to solve. If you follow the steps here you're less likely to make those mistakes. You'll have a big-picture understanding of the scope of problems and so be in a better position to choose the right problem to begin solving. Larry and Sergey are a testament to choosing the right problems and understanding the entire scope of problems that they faced, and like Newton and Einstein, they are able to develop elegant solutions to them.

Of course, you know how the story of the Google guys ends. The algorithm they created to determine the importance of Web pages, based on a link rating system, became known as PageRank (a play on Larry's last name) and was filed with the United States government under patent #6285999 as the intellectual property of Stanford University. Using the search problem matrix they developed in Gates 360 as a foundation for the ideas they would create, they left the university before finishing their Ph.D.'s and moved into a garage a few miles away and started a company on this foundation. Within a few years Google would become the most popular site on the Internet and Larry and Sergey the most influential businessmen of their day. The patent was ultimately sold by Stanford in 2005 for $336 million and Google would become worth nearly $200 billion. Today, if you ever get the chance to tour the campus, make sure that you visit the southwest side. There you'll find a beige building with a red tiled roof and you can look up to the third floor and say, "Yeah, that's where they worked on the *problems of search* in 1996 and 1997. And you know, it *was* a big deal."

• • • •

So, the *First Step* in *Borrowing Brilliance* is to build a foundation of problems. It's this foundation that you'll use to construct your creative ideas, just as Larry and Sergey constructed theirs on a foundation of search problems. While defining the problem is the *First Step*, it's closely tied with the *Second Step*, for you'll use this foundation as a map to finding your solutions. In other words, if you've got a "search" problem, then you'll ask yourself: *Who else has a "search" problem?* Then you'll answer: librarians, rescue teams, sailors, hunters, archaeologists, researchers, and explorers. Then you'll go to these places to look for ideas and solutions to the problems you've defined. That's why author Norman Vincent Peale said, "Every problem has in it the seeds to its own solution." That seed is in how you define your problem, for it determines where you'll go to find a solution. How you define it will determine how you solve it.

This is true for business, the arts, the sciences, literature . . . for any creative domain. In fact, it's true for life itself. How you define your personal problems will determine who you are and the places you'll go. I know, for that's exactly how my journey out of Tempe, Arizona, began. It was determined by how I defined my problems.

The First Days of the Long, Strange Trip

Tortuous thoughts echo in my mind as I sit in my hidden apartment, trying to devise a comeback strategy. The first days of any journey often seem the longest, it's like climbing above the tree line, there's nothing to judge progress by. No miles behind you. Like walking on a treadmill.

However, using the tools in my notebook I begin to think about my problems. I have a lot of them and group them something like this: I'm broke. Overweight. Have no career. And I'm drinking Stoli and Cran for dinner every night. When I sort them, the career problem goes on top and the drinking one on bottom, just the opposite of how Alcoholics Anonymous or Betty Ford would tell you to rank them. Oh, well. So I start at the top, with my highest-level problem. It seems like the weakest link in my chain of problems and solving it may solve these other ones.

I pick up the phone and called Tom Allanson. Tom was the president of the GE Capital Division who had offered me twenty-five million for Preferred Capital the previous year. He'd left GE before my deal had fallen through with the bank, before I had banished myself from Tahoe. Now he was now at Intuit running the TurboTax Division. He was the only one I could think of who might offer me a job outside the finance industry. And I desperately needed to get out of that industry. I told his secretary I was an old friend. I wasn't sure he'd even remember me. Or take my call.

"Hey, buddy," he said.

CHAPTER TWO

THE SECOND STEP—BORROWING

USING AN EXISTING IDEA AS THE
MATERIAL TO CONSTRUCT A NEW IDEA

Traveling back in time fifteen years, I see myself sitting in bed reading. I should be enjoying the book but something's wrong. I have a frustrated look on my face. Outside, a car horn fractures the silence, breaking the serenity of the neighborhood and the composition of the thoughts in my mind. It's the source of my frustration.

At first I think it's a boyfriend calling for Ashley, my seventeen-year-old neighbor, whom I sometimes refer to as Cinderella. But after a minute of incessant honking I conclude that even Cinderella's prince isn't that rude. It must be a car alarm crying wolf. After another minute, I question that, too, get up, walk over to the window, and look outside. It's then that I realize the sound isn't coming from the street but from inside the house. *Huh?* I think. *That's weird.*

I stop and listen. It's coming from downstairs. And it's joined by another familiar sound; a smoke alarm singing along with it in some kind of bizarre harmony. Something's on fire.

Walking downstairs, I see the living room filled with smoke. The kitchen too. I'm still confused. I look at the door to the garage and realize that smoke is billowing into the room from underneath it. *Holy shit,* the car's on fire!

I throw open the door and there's my Jeep Wrangler, top down, and flames dancing across the driver's seat and onto the dashboard. I freeze. Panic. *I gotta fight it.* The car is about to explode and take my entire home with it. *What to do? What to do?*

I look around the garage for something to fight it with. Next to the Jeep is a bucket I use to wash the cars. I grab it, run into the kitchen, and yell at the faucet as it slowly fills the bucket.

Now, before I tell you what happens next, let me remind you— the worst thing you can do to an electrical fire is to throw water on it. You might as well toss a can of gasoline or lighter fluid on it. Water *conducts* electricity, so throwing water on an electrical fire is like dropping napalm from the sky.

However, in the heat of battle, I fail to recognize this is an electrical fire, even though the garage has the distinct smell of burnt hair. I dump the pail of water on the fire.

A mini-Hiroshima bomb blast follows, as if I've just dropped *Little Boy* on the front seat of my Wrangler. A small mushroom cloud forms, licks the ceiling, and scours the walls of the garage. It's quite beautiful from a pyrotechnic point of view but, from a homeowner's, terrifying. I am thrown back against the wall. My small car fire is now a raging garage inferno, on the cusp of being out of control. I have a full tank of gas, so it's only a matter of minutes before that explodes and, I assume, takes me out, my home, and probably Cinderella as well. I freeze again. Panicked.

A moment later I compose myself. OK. Water doesn't work. At least now I know what I'm dealing with. If water isn't the right material, what is? I recall that a fire needs two things: fuel and oxygen. My

Jeep and I, unfortunately, have provided the former. So I need to starve the fire of oxygen. I need to smother it. I look around the garage, again, and in the far corner is a pile of blankets I had forgotten to return to U-Haul after moving into my house.

I grab a blanket and throw one into the car. At first the flame falters, but then I realize the blanket is on fire and I've just added more fuel to the inferno. Apparently, the U-Haul moving blankets are made of some kind of highly flammable fabric. However, it does seem to have staggered the flame for a moment. This makes me remember that there's a heavy wool blanket in my closet in the living room, and so I run back into the house to get it. Perhaps it won't be so combustible. I throw the much heavier wool blanket onto the blaze. It seems to smother the flames. The fire is knocked over. I use the blanket like a sword, attacking the various pockets of fire and flames as they try and escape my onslaught. Few more stabs at it and fire is out.

The garage is filled with smoke and the smell of burnt wire and plastic. I open the door to let the smoke escape to the street. The car horn is still blaring, and a smoke alarm is still screaming in harmony along with it. I stumble out and sit down. Behind me neighbors have gathered to watch the ridiculous spectacle. I turn around and see Ashley looking down at me, shaking her head in disgust. *Take a hike, Cinderella,* I think. *You have no idea how close you just came to being blown to fugging pieces.*

• • • •

Fifteen years later, as this story pops into my head, I realize there are two things to be learned. The first is to read my mail, for I found out later my car had been recalled, the notice still on my desk, unopened. Apparently, my Jeep Wrangler wasn't the first to have experienced spontaneous combustion. The second is that the materials I choose to solve my problem will greatly determine the quality of that solution. Use the wrong material and I run the risk of blowing up my entire home. Sometimes you have to venture far away from your problem in order to solve it—as I had to venture into the house, away from the garage, in order to solve my fire problem. Later I'd learn that wool is a naturally flame-retardant material.

Go figure, right?

• • • •

The Source of Ideas

Evidently, Stephen King and his fellow writers fear asking where ideas come from, for in his book *On Writing* he says that he, Dave Barry, Ridley Pearson, and Amy Tan are ". . . writers, and we never ask one another where we get our ideas; we know we don't know."

However, within ten pages, he tells the reader about the first story he ever wrote, in first grade, at the age of six. He says, "Imitation preceded creation; I would copy *Combat Casey* comics word for word in my Blue Horse tablet, sometimes adding my own descriptions where they seemed appropriate." In other words, first he copied, then he created. *Sound familiar?* He admits he began his career as a plagiarist. He took someone else's words, someone else's ideas, changed a word here and there, and then called the copied passage his own. So, at this point in his career, he knew exactly where he got his ideas from—he got them from someone else.

He says that now it's different, for ". . . good story ideas seem to come quite literally from nowhere, sailing at you right out of the empty sky." Yet, he explains how the protagonist Carrie White in his first novel, *Carrie,* was a combination of Sondra and Dodie, two "real" girls he knew in high school. For the next few pages he relates stories about these girls, stories that became episodes in his novel and scenes in Brian De Palma's movie adaptation. What he's telling us, quite literally, is that he took his life experiences, stole these ideas, borrowed them, combined them with other things, and built something new and creative out of them. This is not unlike how he created his first story, the one he copied nearly word for word. Ironically, he describes where he gets his ideas from in the same chapter he tells us his ideas come from "nowhere." Do King and his fellow authors fail to ask where they get their ideas from because they "don't know" or because they don't want to admit their source? You see, Stephen and his fellow writers borrow their ideas just like everyone else. It doesn't matter if you're trying to solve a business problem or a character development problem: First you copy, then you create. All brilliance is borrowed.

I don't mean any disrespect by this; in fact, Stephen King is my favorite novelist. I consider him a creative genius. I read *Misery* in one sitting, never getting up from the couch once I began. I was going to name my daughter, Katie, after him if she had been a boy. And I don't think he's lying about his ideas, I think they do seem to come from nowhere and out of the empty sky. Later, in the "Incubation" chapter, I'll explain how ideas form in the subconscious and then appear when you least expect them. I'll also explain that those ideas, while seeming to be unique, are creative combinations of borrowed ideas constructed in the shadows of your mind, away from the light of conscious thought. In addition in that chapter, I'll teach you how to facilitate this subconscious process. In the meantime, you need to learn how to borrow the right stuff, and how to feed the subconscious mind with the material it'll need to construct its solutions. For the materials will determine the design.

Let me explain.

Materials Determine Design

In 1989, I was named senior manager of advanced technologies for the McDonnell Douglas Astronautics Company, the largest defense and aerospace contractor in the world at the time. This sent me from California to Washington to work closely with the Defense Advanced Research Projects Agency (DARPA). The agency was established in 1958 after the emotional shock of the Soviet Union's launch of Sputnik, the world's first orbiting satellite. DARPA was chartered to develop space age technologies and to prevent the United States from ever being "Sputniked" again. I spent a year working with the agency in its nondescript building in Rosslyn, Virginia, expecting to get involved in highly complex and highly secretive advanced weapon systems like a neutron particle beam ray-gun or an army of android robotic warriors. What I found, instead, and much to my disappointment, was the majority of the advanced research was "materials" research. Most of what I worked on was not to build an army of automated pig-man soldiers, but advances in alloys, adhesives, ceramics, and sophisticated fiber resin composites like Kevlar. What I realized, and what you must, is that the construction material determines the design of whatever is to be built, whether a bridge, a building, a ballistic missile, or a robotic pig-man warrior.

Historians define human time periods in terms of the primary material of each epoch: the Stone Age, the Bronze Age, and the Iron Age. Some call our time the Silicon Age. The structures, weapons, and tools from each period were dependent upon the materials, for the materials determined the designs of the era. It was impossible to design a sword in the Stone Age—the material wouldn't allow for it—so swords don't appear in the archaeological record until the Bronze Age.

The Wright Brothers couldn't have conceived the design of a hypersonic stealth aircraft because they worked with wood, canvas, and wire. It wasn't until DARPA perfected dielectric composites and radar absorbent material that designers could imagine an invisible aircraft flying undetected at twice the speed of sound. If Orville came back today and saw a modern stealth fighter sitting on the tarmac, he might not recognize it as an airplane, although he was a gifted aircraft architect. And although the plane's function remains the same, the materials have completely changed the form, the structure, of the object itself.

Just as the materials engineers use determine the design of their creations, so, too, will your ideas be influenced by the materials you use to construct them. Before constructing, you gather the materials for your solution, just as I did in North Conway before building my raft, or in my garage as my car went up in flames. In each case the materials determined the nature of the solution. You can't make something out of nothing; you have to make it out of something else. Every chef knows the ingredients in your kitchen will determine the recipe and dinner you serve your guests. Where you get these ingredients from, the places you shop, determine the opinion you and others have of your new idea.

Let me explain.

Source Determines Perception

Since ideas are constructed out of borrowed or stolen conceptions, the source of these thefts will determine the idea's perceived "degree of creativity." As Albert Einstein said, "The secret to creativity is knowing how to hide your sources." The point for you, in this step, isn't how to hide your sources but to determine them (you'll learn to hide them later). To do this, it helps to understand how the source determines this perception, or how where you shop determines the uniqueness of your meal.

Imagine, for a moment, yourself as a software designer gathering ideas to create a new tax-software program. Now imagine putting these sources on a continuum. At the far left of the continuum, as shown in Figure 1, are the places from within your domain. At the far right are the places outside your domain, far away and not usually associated with tax software.

You begin gathering ideas by going to TurboTax, for you like the way it uses a single question on each screen in order to gather tax data. You borrow this and do the same thing with your design. Since TurboTax is a competitor, you are perceived as "stealing" the idea. You're a thief.

FIGURE 1: PERCEPTION VS. SOURCE

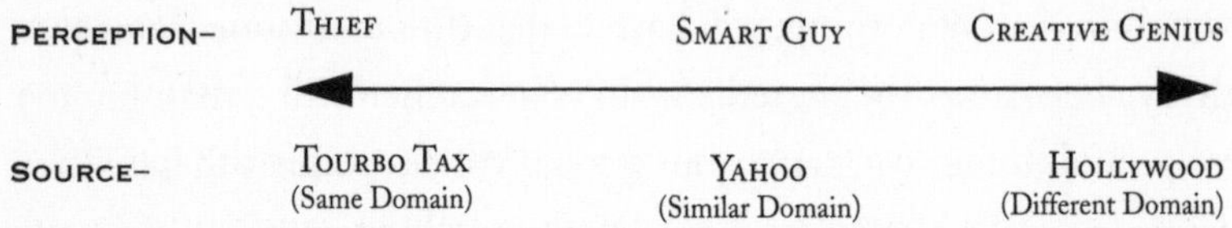

PERCEPTION–	THIEF	SMART GUY	CREATIVE GENIUS
	◄――――――――――――――――――――――――►		
SOURCE–	TOURBO TAX (Same Domain)	YAHOO (Similar Domain)	HOLLYWOOD (Different Domain)

Next you go to Yahoo!, study the site, and like the way the Help Center is laid out. You borrow the idea and do the same thing with your design. Since Yahoo! is not a direct competitor, your idea is perceived as fairly creative, and you've hidden your source, as Einstein suggested. Now you're a "smart guy."

Finally, you go to Hollywood and visit with an Academy Award–winning director to learn how to create an emotional response in an audience, a problem you'd like to solve with your product. He explains every Hollywood movie is structured as a three-act play. In the first act, the first ten minutes of the movie, a conflict is established between the hero and the villain. In the second act, the next hour and a half, the hero tries to resolve the conflict, only to be prevented by the villain. In the final act, the climax, the last thirty minutes of the movie, the hero resolves the conflict. You like this structure and borrow the idea and do the same thing with your tax software. You begin the program by defining a conflict. The first screen reads: "Every week the IRS has taken money from your paycheck. Now it's your turn to take it back!" You establish the user as the hero and the IRS as the villain. You follow the Hollywood pattern throughout the program until the end, when the user overcomes the conflict and gets the maximum refund, thwarting the villain. Since you've borrowed this idea from a place far removed from your domain, at the far right of the continuum, you are perceived as a "creative genius."

Ironically, whether you're a thief, a smart guy, or a creative genius, it doesn't matter, you're still doing the same thing: You're borrowing your ideas and combining them with other things to construct your creation. What matters is where you take the ideas from. The source determines the creative perception of your solution. After all, any schmuck can throw water on an electrical fire from the garage; it takes a smarter schmuck to venture into the house and find a flame-retardant wool blanket to smother it. The farther away from your subject you borrow materials from, the more creative your solution becomes. The ingredients used determine the originality of the banquet. Borrow exotic materials and you can expect an exotic feast. Any biologist can find solutions in biology, the creative biologist finds them in astronomy. The creative astronomer finds her ideas in macroeconomics. And the creative economist finds them in a Saturday-night poker game—just ask John Nash, the Nobel Prize–winning mathematician depicted in the movie *A Beautiful Mind*. First you copy. Then you create. There's a fine line between plagiarism and creativity, a line defined by the source of the theft.

You're not quite ready to create, first you have to copy. As T. S. Eliot said, "Immature poets imitate; mature poets steal."

The Search for Materials

Let's get started. Since a new idea is constructed out of existing ideas, the second step in borrowing brilliance is the search for these ideas.

The search, like any journey, begins close to home: You borrow from within your domain by studying your subject, your competition, and making observations. Then you take a step away and study those subjects similar to yours, looking for similar problems and their solutions. Finally, you travel to faraway places, exotic places never associated with your subject. From there you borrow these exotic materials and use them and the others to begin constructing your solution. Ultimately, if you're a businessperson, you'll combine the materials borrowed from your direct competitors with the materials borrowed from other industries, with the more exotic materials collected from places unrelated to business, such as science, nature, or entertainment, and use this stuff to construct a truly creative business idea. If you're a chemist, you'll combine borrowed materials from other chemists, with the materials borrowed from other scientific domains, with the more exotic materials from places unrelated to chemistry and science, such as business, economics, or politics, and use this stuff to construct a truly creative chemical idea.

Thankfully, this isn't a blind search, for you've already constructed a map that'll lead you on this journey. Your foundation from the previous chapter serves as a search grid for your borrowed solutions. In other words, you'll look to places, whether near or far, that are tied together by problems similar to yours. For example, if you're constructing a new piece of software and you've got a "navigation" problem—users getting lost—then you'd ask yourself: *Who else has navigational problems?* First, you'd go to other software designers to borrow navigational solutions. Then you'd go to similar industries, like video game products, and borrow navigational ideas from them. Finally, you'd travel to those places or people not associated with software, but with their own navigational problems, people like sailors, pilots, truck drivers, explorers, and rats caught in a maze. Once you've got all these materials, then you'd begin constructing your solution by making unique combinations of this stuff.

The creative thinker travels near and far in search of solutions. If he never ventures too far from home, if his ideas are borrowed from within his own industry and not combined with more exotic borrowings, he is regarded as a thief and his conceptions as pilfered, even if they're very successful solutions. Bill Gates became the richest man in the world by solving important software problems, but most of his creative solutions were borrowed from within the software industry itself. This earned him his reputation as one of the pirates of Silicon Valley. Charles Darwin, on the other hand, became a revered scientist by solving important biological problems but he was never called the pirate of the Galápagos because his creative solutions were borrowed from places not usually associated with biology. Gates never traveled far from home to solve his problems. Darwin did. Both were very effective thinkers and both changed the world with their creations, and so there's something important to be learned from each: how to solve problems by borrowing solutions from others and how the place you borrow from changes the creative assessment.

Think of your journey the way a search-and-rescue team does. Thomas Bayes, an eighteenth-century British mathematician, developed a probability theorem used today by professional rescue teams. Using Bayesian theory, a team identifies variables like the last known position of the lost thing, its average speed, its direction, and other known variables. From this data, they construct a search path, using statistics, which starts at the point of highest probability, then moves on to places of intermediate probability, and then finally to places of low probability. The U.S. Navy used this method to find a lost hydrogen bomb in 1966 when a B-52 bomber collided, in midair, with a KC-135 refueling plane and dropped off radar near the coast of Spain. Your search for ideas should use the same process. Start in places of high probability, places you're most likely to find the right material; and then work toward places of lower probability. The place of highest probability is usually within your own industry, so always begin the search there, since people in your own industry are most likely to be solving the same problems.

Borrowing from Competitors

When you borrow from someplace close, from your competitors, you're often considered a lowly pirate. However, borrow from outside your industry and you're considered a creative genius. The historical pirates of the Caribbean, like Englishman Henry Morgan, were not considered bandits but soldiers of fortune because they stole exclusively from Spanish colonies and not British ones. Captain Morgan was called a "privateer" in England, hailed as a hero, and given important political appointments by Queen Elizabeth. It wasn't until years later, when his men started stealing from British colonies and British merchant marines, that they were considered thieves and outlaws. The privateers became pirates and they were hunted by the same people who had sanctioned their trade in the first place. They'd been taking all along but had been taking from foreigners, not from their own countrymen. The source of the theft changed how it was perceived.

Intellectual borrowing works the same way. Borrow from within your industry and you are considered a thief or lowly pirate. Borrow from another industry and you're considered a hero and a creative privateer. Does this mean you stay away from the ideas of your competitors? Certainly not, just don't take exclusively from them. Mix it up. Do lots of borrowing to cover your tracks. As screenwriter Wilson Mizner noted, "If you steal from one author, it's plagiarism; if you steal from many, it's research." Do lots of taking and you'll be fine. Your borrowings will be lost in the combinatory construction and hard to detect. As Einstein said, you'll learn to hide your sources. This happens through the natural evolution of your ideas. The thefts get buried deep in the DNA of your solutions and are not on the surface for all to see.

The creative journey for Bill Gates began in the early seventies very close to home, in a makeshift computer laboratory at Lakeside High School in the Seattle suburbs. At the lab Gates became fascinated with the emerging world of computers. He spent endless hours hacking away at a teletype machine connected by a phone to a PDP-10 computer located several miles away at a company called Computer Center Corporation. It was also at the lab that he met Paul Allen, two years older, and another hacker fascinated with these thinking machines. Seven years later they would form a company to write software that would become the most powerful and successful start-up in the history of American business.

In those days, computing time was a valuable asset and the school was billed by Computer Center Corporation for the amount of time the students used the machines. Pretty soon Gates, Allen, and others were racking up thousands of dollars in bills. School officials told them to stop. Instead of slowing down, Gates hacked into the central computer, found the accounting files, and erased most of the hours on his personal account. He was proud of this until he was found out. The manager at the computer center drove to the school and in the principal's office confronted the young pirate. As punishment, Gates couldn't use the computer for six weeks. However, the manager was so impressed with the young hacker that he offered Gates a job finding "bugs" in the center's software. Gates proved to be extremely good at this and was paid, ironically, in computer time. From the beginning, Gates had a knack for intellectual borrowing and he'd use it throughout his career. He'd perfect it over time the way an artist perfects his brushstroke. Later, it would earn him a place as the captain of an intellectual pirate ship competing with the other pirates on the seas of the burgeoning personal computer industry.

The inception for Microsoft began while Gates was attending Harvard. A small company in Albuquerque called MITS (Model Instrumentation Telemetry Systems) created the first personal computer, called the Altair 8800. Paul Allen read about the primitive computer in the January 1975 issue of *Popular Electronics* and showed the article to his old high school friend. Much to their credit, they both recognized the PC revolution the day it started. However, the Altair was a crude machine. The operator flipped switches on the front panel to make it work. It had no printer, no monitor, and no software to control it.

Gates saw his opportunity—develop the software for this computer and the ones he knew would follow in its wake. So, in the first of many marathon code-writing sessions, Gates and Allen constructed the first software language for the Altair over the next eight weeks. They borrowed the structure of the language from the one they'd used on the teletype machines in the Lakeside computer laboratory. Called BASIC, it had been developed by a Dartmouth professor and was in the public domain. Gates and Allen copied it and then adapted it to run on the Intel 8080 chip that controlled the Altair computer. First they copied, then they created.

A few months later, Allen flew to Albuquerque and presented the software to Ed Roberts, owner of MITS. According to Allen, he fed his program into the computer through a paper tape reader connected to the machine. He crossed his fingers, for it was the first time he'd ever touched the Altair: he and Gates had developed the software on a different machine in the Harvard computer lab. When the Altair was ready, Allen typed: "Print 2 + 2." The Altair hesitated for a moment, and then printed out "4."

"Those guys were really stunned to see their computer work," Allen said. "This was a fly-by-night computer company. I was pretty stunned myself that it worked the first time. But I tried not to show much surprise."

Gates and Allen would become the first software provider for the first personal computer ever produced. Paul moved to Albuquerque and a few months later Bill finished his sophomore year and joined him. Like Larry and Sergey, he'd never return to complete his studies. They formed a corporation called Micro-Soft and split the stock 60/40 in favor of Gates, since he argued that he had done most of the initial coding for the Altair BASIC program. Paul Allen didn't seem to care at the time.

From the beginning Gates was focused on one very well-defined problem: dominating the software market for the personal computer. He'd say, "A computer on every desktop and our software in every computer." According to Eddie Curry, one of the developers who worked on the Altair, Gates and Allen "... had a very clear understanding of what they were doing, in a sense that they had a vision of where they were going. It wasn't just that they were developing BASIC. I don't think that most people ever really understood this, but Bill, certainly, always had the vision from the time that I met him that Microsoft's mission in life was to provide all the software for microcomputers."

In 1981, IBM entered the personal computer market. At the time, IBM's name was synonymous with computers—the IBM 360 was the standard by which all computers were measured. At first, the company felt little competition from the personal computer because most were being used by hobbyists and not businesses. Then Apple changed that with the Apple II, which ran a clever new program called VisiCalc, the first spreadsheet. Businesses bought Apples in bunches. IBM had to play catch-up, so they outsourced many of the components of their new machine, including, of course, the software.

A computer uses three levels of software. The first level contains the application layer—the user interface, which includes things like spreadsheets and word processors. The second level holds programming language like BASIC or FORTRAN. Coders use this language to write the applications. The third level is the operating system, like DOS or WINDOWS. This is the interface between the application and the binary world of machine language. The operating system runs the tasks of the computer, such as finding disk space or storing files. A machine cannot run without an operating system.

IBM contacted Bill Gates and negotiated with him to purchase MS-BASIC. At first the company wanted to own it outright, but Gates was adamant about making it a licensing or royalty deal. He would let IBM borrow it, but not let them take it away from him. Then IBM approached a company called Digital Research to buy an operating system, a product Microsoft didn't offer. Unable to strike a deal with Digital Research, IBM asked Gates if he could negotiate for the operating system on its behalf.

Well, negotiate he did. Sensing a business opportunity, instead of going back to Digital Research, Gates went to a small company called Seattle Computer Products, which had copied the PC operating system concept from Digital Research and developed its own simplified version of it called QDOS (Quick Disk Operating System). Gates offered to license the code from Seattle Computers for fifty thousand dollars, a very nice sum for the small business. Of course, Gates forgot to mention that he was going to turn around and sell it to IBM. Seattle Computers quickly agreed to the deal because it needed cash. However, when it came time to sign the papers, Gates changed the contract. Instead of Microsoft licensing QDOS from Seattle Computers, the contract was worded so Microsoft would own QDOS and license it back to Seattle Computer. Seattle Computer signed the contract not understanding the monumental impact the new wording implied. The rest, as they say, is history. Gates renamed the product MS-DOS, improved it, and freely licensed it to IBM under another lucrative royalty agreement.

Bill Gates had pulled off the business deal of the century. IBM would sell millions of PCs, each running MS-DOS, and each triggering a royalty check to Microsoft. Others would copy, or clone, IBM's machine and they, too, would turn to Gates for his borrowed operating system. Bill's vision of a computer on every desktop and Microsoft inside every computer slowly became a reality. Gates had borrowed the code from Seattle Computer, which had borrowed it from Digital Research, and used it as a beachhead into the desktops of millions of computers, brilliantly solving the problem he had identified. What he had outright purchased for only fifty thousand dollars was, within a few years, bringing in over two hundred million dollars in revenues and would bring billions more in the future. The business deal of the century made him the richest man in the world and for us is the perfect example of what I mean by the term *borrowed brilliance*.

However, Bill Gates isn't the only one to successfully borrow ideas from competitors, he's just the most successful one. In fact, in business, competitive borrowing is an important strategy recognized by most successful business leaders. According to marketing experts Al Ries and Jack Trout, in the book *Marketing Warfare*, competitive borrowing should be the primary strategy for every market share leader of every industry. Successful companies should watch the marketplace and look for new ideas emerging from the smaller players. Then, either copy the new idea or buy the company if it's a patentable innovation. New product categories are won by the company first to dominate that category, as Bill Gates and Paul Allen did.

Market leaders are in the best position to execute a dominating tactic—after all, they have the resources. Those that refuse to copy, or copy too late, lose their market positions, as IBM would lose its position to upstarts like Apple and Microsoft.

For example, Gillette has been an innovative leader since its founding by King Gillette in 1901. Gillette borrowed ideas from within the shaving and razor blade industry, improved on them, and went to other industries to find things to combine with his borrowings. His successors have continued his creative approach. For example, BIC, the disposable-pen company, entered Gillette's market by creating the disposable razor in 1975. However, Gillette quickly copied it, improved it by adding an extra blade, called it the Good News Razor, and used its distribution channels to dominate the market before BIC could effectively promote its own innovation. Gillette recognized a good idea and seized it. Today it still maintains an 80 percent share of the market. Are its executives considered pirates? I bet they are by BIC. But they've borrowed from so many different places, improved the borrowings, and so covered their creative tracks.

Borrowing from within your subject, close to home, isn't just for businesses. Copying is used effectively in every domain. Michelangelo copied techniques from Leonardo da Vinci. Hemingway borrowed important ideas, and even words, from other writers. The title *For Whom the Bell Tolls* is a line stolen from a John Donne poem. *The Sun Also Rises* is almost a direct quote from the King James Translation of the Bible. Isaac Newton borrowed ideas from the other mathematicians like John Wallis and René Descartes to create calculus. He cryptically said, when accused of plagiarism by another mathematician, that he had "stood on the shoulders of giants." Of course, Quentin Tarantino, the writer and director of *Pulp Fiction,* wasn't as cryptic when he said, "I steal from every movie I've ever seen."

I don't think it's fair, though, to believe these creative thinkers are thieves, just as I don't think it's fair to believe Bill Gates is an intellectual pirate. Robbery and theft imply you have taken something away from someone else. Sure, Gates took his ideas for an operating system, but he never stopped Seattle Computer or Digital Research from using these conceptions. Both were free to market and sell the concepts Gates had legally borrowed. They just didn't do it as effectively as Gates did. Gates was merely doing what every creative thinker has done since the beginning of time. Today, the same person whom people accused of intellectual theft is taking the billions of dollars he acquired through creative thought and giving it back to society through a massive charitable foundation in a move reminiscent of, or borrowed from, Andrew Carnegie and John D. Rockefeller. Not exactly the move you'd expect from a real pirate, now, is it?

With that said, there are other ways of borrowing from within your domain. The most popular and effective is simply borrowing from your own experiences in your subject area.

Borrowing from Observations

While copying others is an extremely valuable ploy, it's one often frowned upon by creativity purists. It is (unfairly, in my opinion) considered cheating. Observation, on the other hand, is considered perfectly acceptable by the same critics even though, technically, it's the same thing. You're just taking from what you observe, instead of taking from an associate, competitor, or business partner.

With that said, observation does, as scientific philosopher Karl Popper noted, provide the source for all knowledge. It's the backbone of the scientific method. It's how this crazy process of creativity began in the first place. In the beginning of time, there were no ideas to steal, no competitors, no intellectual conceptions, only the natural world unfolding in a natural way. And while no one recorded the invention of the wheel, as I said earlier, I suspect it happened like this . . . a Neanderthal was hiking in the hills behind his cave, dislodged a rock, and watched it roll down the slope, and he went "aha." He borrowed this idea and chiseled the first wheel out of a stone and then took it up to the top of the same hill and let it roll down under its own weight, amazing his family and cave-dwelling friends. The endless processing and evolution of ideas began from this simple observation.

Of course, you need to continue this time-honored tradition of borrowing observations. As you go out on your search for new ideas, keep your eyes open, hone your observational skills. Observe your customers. Look for the making and breaking of patterns. This is the same tool I described in the last chapter, the same tool we used to identify problems. Just as a thief can use a wire cutter to do two different things, like disable an alarm and cut a hole in the fence, you can use the observation tool for both identifying problems and borrowing solutions. Remember how Scott Cook had established the "Follow Me Home" program at Intuit to observe customers? He was observing both problems and solutions. Customers were using a personal finance program to solve business problems, an observation Cook and others would borrow and use to develop QuickBooks and turn it into a billion-dollar business.

It's not just customers but the observation of anything that can provide the inspiration, the material, for constructing a new idea. For example, the design of the Empire State Building came from a very simple observation made by an architect while he sat at his drawing board. The building itself was conceived in 1929 by Jakob Raskob, the founder of General Motors. He was involved in an intense business and personal competition with Walter Chrysler, the founder of Chrysler Motors. When Chrysler broke ground in midtown Manhattan to erect the tallest building in the world, he took Raskob by surprise. So Raskob hired the architectural firm of Shreve, Lamb & Harmon to design a skyscraper even bigger and bolder than Chrysler's. Sensing a game of catch-up, Raskob gave the firm a month to complete the architectural plans. Under intense creative pressure, the architect William Lamb sat in front of a blank piece of paper, wondering how he was going to design such an important building in such a short amount of time. On his desk was a simple #2 pencil. At some point, he picked it up, held it at arm's length, and curiously admired its shape. And so, as legend has it, the design for one of the great architectural and engineering accomplishments of the twentieth century was born from the simple observation of the shape of a #2 pencil.

Writers of fiction and nonfiction use this tool to great effect. Ken Kesey's story *One Flew over the Cuckoo's Nest* was borrowed from his own observations as a male nurse at a psychiatric hospital. The screenplay *Casablanca* was borrowed by Murray Burnett from a 1938 trip he took to Vienna shortly after the annexation of Austria by the Germans. He observed the bizarre situation of Nazis coexisting, peacefully, with allies and refugees in other parts of the world. *The Sun Also Rises* borrows Hemingway's pilgrimage to Pamplona and his observations of the annual fiesta and the running of the bulls. These personal observations provide the source materials the writer needs to construct a creative story. The writer borrows from the people, places, and things he observes.

Borrowing from Other People

In addition to borrowing from competitors and observations, the creative thinker borrows from the people around him, within his own organization or from his business partners. A successful executive surrounds himself with creative thinkers, borrowing ideas and using them to solve important business problems. As product life cycles become shorter and businesses are forced to reinvent themselves and their business models, this reliance on employee innovation becomes more important. So, you need to learn how to take ideas from the people around you. Ideation is a collaborative effort.

Thomas Edison said, "I readily absorb ideas from every source, frequently starting where the last person left off." Most of his inventions were collaborative efforts, as he borrowed the ideas of his fellow inventors at his laboratory in Menlo Park, New Jersey. Abraham Lincoln remarked, "I can learn something from every man that I meet." And Newton, I repeat, stood on the shoulders of giants. Each of them gathered material by going to others for help as a regular course of business.

Bill Gates spent a lot of time and money to hire the most brilliant programmers in the world as he built his company. He wasn't afraid to borrow the solutions constructed by his employees. He said, "At Microsoft there are lots of brilliant ideas, but the image is that they all come from the top—I'm afraid that's not quite right." He borrowed ideas from those around him. Not unlike the other pirates of Silicon Valley. His nemesis, Steve Jobs, is notorious for taking the ideas of others. I've never worked with Steve, but being in the software industry myself I have friends and colleagues who have. No one denies his immense creative abilities. They do, however, each tell a similar story about how he borrows ideas. Don Denman, a young programmer who worked with Steve on the Macintosh, said, "We all had a joke about Steve. If you want to get him to agree to a new idea, something that was a good idea but that he hadn't thought of, you told him the idea, and then just let him reject it. A couple of weeks later he would come rushing over to you and tell you how he had just had a great idea and would proceed to tell you the same idea you had told him before."

In fact, the most blatant case of borrowed brilliance in business history involved the battle between the two most celebrated pirates of the valley: Jobs and Gates. You see, Gates and Jobs began as business partners, not competitors. Their partnership would culminate in the infamous lawsuit between Apple and Microsoft.

At the same time Gates was working with IBM to develop the PC, he was partnered with Apple to develop software for Jobs's newest product, which would eventually become the Macintosh. Before it was released, Steve showed it to Bill, since he wanted Bill to develop a version of Microsoft BASIC for the new machine. This computer would be the first commercial product to use a mouse and graphical user interface (GUI), and it took Gates completely by surprise. He instantly saw the value of the mouse and GUI. When Gates returned to Redmond he began working on the next generation of the MS-DOS operating system and incorporating this new technology. A few years later, when Gates released the software, Jobs became infuriated, realizing that Gates had taken his idea and released the new product under the cleverly marketed name *Windows*.

Apple cried foul. "You've stolen our idea!" they exclaimed. Of course, this was clearly the pot calling the kettle black. The Macintosh itself, its original conception, was taken by Jobs after a visit to the Xerox Palo Alto Research Center. In fact, Jobs hired the Xerox engineer who developed the mouse and GUI and put him to work on the Mac. And Xerox, ironically, had taken the technology from an obscure academic researcher. Jobs was well aware of where his ideas were coming from; as a matter of fact, the Cupertino building in which the Mac was developed had a pirate flag ceremoniously pinned to a wall in its conference room. Jobs would say of Apple, "Why join the navy if you can be a pirate?"

In 1988, Apple Computer filed a copyright infringement lawsuit against Microsoft. The suit claimed Microsoft had stolen visual display features from the Mac and used them in Windows. What the suit failed to mention was that Apple had taken the concept from Xerox. Dan Bricklin, a prominent software engineer who developed the first spreadsheet, said of the suit, "This is a sad day for the software industry in America." He added, "Writing software is not the same as writing a book. Software builds on what was there before." Five years and millions of wasted dollars later, Microsoft won the suit. The company declared victory for themselves and all creative thinkers. Bricklin was right. Software builds on what was there before, as does every commercial product, engineered machine, scientific theory, and creative thought. The only thing Bricklin was wrong about was that books, and the ideas in them, are also built on what was there before. Books are built out of other books—that's why nonfiction books have a bibliography. Language, after all, is borrowed too.

For some, this is dirty business. For others, it's business as usual. For me, the difference between what Gates, Jobs, and the other pirates of Silicon Valley do and what Newton, Hemingway, and Picasso do is in the source of copied material, not in the act of copying. Remember, creative thinking is problem solving, and you find your solution in places with a similar problem to yours. Well, your competitors have a similar problem, so you're obligated to look there. But, if you're not comfortable flying a pirate flag in your conference room, then you'll need to venture away from home, away from your industry, and borrow from foreign lands, other industries, other places, and not from your own industry. You'll have to go outside the garage to solve your fire problem. Like Captain Morgan, you'll act like a highly regarded privateer and not like an ill-regarded pirate. You're taking ideas, but not from your countrymen. It's a bit more challenging but it's a lot more rewarding. Is it more effective? Perhaps, but not necessarily. You can't argue with the success of the Silicon Valley pirates. Borrowing from faraway places will, however, result in more "creative" solutions, just as borrowing from a Hollywood movie to construct a tax software program is more "creative" than borrowing from TurboTax.

No one epitomizes intellectual privateering more than Charles Darwin. Like Gates, he borrowed his ideas, but from places so far away that he was awarded the deep respect of his fellow countrymen for his mastery at borrowing brilliance.

Traveling Away from Home

Darwin began his creative journey on board the HMS *Beagle* in the spring of 1831. The *Beagle* was chartered to travel to faraway places, exploring and mapping the southern hemisphere for the British navy and merchant marines. Darwin was just out of college; his job was to provide intellectual companionship to Captain Robert FitzRoy and his officers on the two-year excursion. This was a common practice in the British navy, since years at sea were tedious and the ship usually manned by uneducated, uncouth, and unscrupulous deckhands. The highly educated FitzRoy wanted the scholarly stimulation that a recent college graduate could provide. As luck would have it, though, the ship's field biologist abandoned his post because he couldn't stomach the idea of being on board for years. He left at the first port-of-call. So FitzRoy assigned his duties to the young and inexperienced Darwin.

While the captain and crew managed the ship, Darwin collected biological specimens by dragging a plankton net behind the *Beagle*. He made long excursions ashore, took copious notes, all the while observing and collecting the plants and animals he discovered along the way. The ship made continental stops in South Africa, Brazil, Uruguay, Argentina, and dozens of other places. They visited the island archipelagos of Tahiti, the Azores, and the Galápagos. A two-year voyage slowly stretched into nearly five years.

During this time, Darwin immersed himself in the natural world. Like Gates plunging himself into the new domain of software in his high school computer laboratory, Darwin was engrossing himself in the new domain of field biology in his Galápagos laboratory. His predecessor had left biology books on board and Darwin sent word to England to have others forwarded to the faraway ports so he could learn from the most prominent biological thinkers of the day.

At first, Darwin copied the ideas of others in his domain, just as Gates and Jobs would do. Then he began using his observations to further his understanding of his adopted subject. You can see the blatant borrowing in his early letters home, many of which were published in the local papers. But then he did something different. As the *Beagle* traveled farther and farther away from her native soil, Darwin's thoughts began to travel farther and farther away from their domain, seeking new ideas in places that no one else had explored, just like the ship he was sailing upon.

Borrowing from the Opposite Place

When looking for borrowed ideas, your first step away from your industry should be in the opposite direction. If success in your market is in making big things, then try making small things. If your success is in making soft things, then consider making hard things. To adopt the opposite of a popular idea is always a novel approach. So go to the opposite place and gather its material. This is one of my favorite thinking techniques because it's simple to use and yet is perceived as being extremely creative. If you take someone's idea, but then disguise this borrowing by using its opposite, they'll call you a creative genius.

Every idea has an opposite. In fact, you can't define something without implying an opposite. There's no concept of light without the concept of darkness. No success without failure. No heights without depths. Opposites characterize each other and that's why they're called counterparts. Opposites attract because they're so closely related.

Darwin's first intellectual step away from the accepted thinking of his domain was a simple about-face. Instead of being an academic pirate and taking the same intellectual position as other field biologists, he decided to take a contrary position. You see, while others were traveling the world and sorting and cataloging different species by noting the subtle differences among various plants and animals, Darwin did the opposite and began cataloging their similarities. For example, he was intrigued by the birds on San Cristóbal in the Galápagos and how they were remarkably similar to those he had seen a year earlier in the Azores thousands of miles away. Sure, there were subtle differences, but he was more fascinated by the resemblances than he was by the variations. Birds were birds no matter where he went. In fact, he even wondered why the wing of the finch was remarkably similar, from a structural standpoint, to his own hand. If you studied the skeletal form of a bird's wing, it resembled a macabre hand with elongated fingers. This thinking was the opposite of the thinking of the others in his field. And it was this thinking, while he was still on board the *Beagle,* that began to convince him that these animals were not different species that had originated in different ways, as was thought, but that they were related. He began to believe that the finches in the Galápagos and those in the Azores were descendants of the same birds, somehow separated, evolving and adapting to different environments. This accounted for the subtle differences. In fact, he reasoned, he could see similarities among all species and began to think that maybe all living things derived this way. Maybe he and the finch had a common ancestor. This was radical thinking, especially in the biblically oriented times of the Victorian era in which he lived. It was opposite thinking. He borrowed the thinking of his fellow biologist but then took a step away by taking the contrary position.

Business history is filled with stories of companies doing the opposite of the market leader and becoming market leaders themselves. For instance, the cola wars were defined by this type of thinking. Coca-Cola, the market leader, had established the category in 1886 and quickly became the most popular drink in the country. Others tried to copy the idea but with little success. For the next three decades Coke dominated the industry with an 80 percent market share, destroying its competition. Then, in 1923, the Loft Candy Company bought the bankrupt assets of one of these competitors, a company called Pepsi-Cola.

The president of Loft, Charles Guth, struggled to make the product profitable, but he was no match for the Coca-Cola Company. It wasn't until Guth accepted the dominance of his competitor and decided to do the opposite that he became competitive. If Coke had been the preferred Cola for the last two generations, Guth reasoned, then let's be the opposite, let's target the younger generation. He turned this strategy into the tactics that would appeal to them by adding more sugar, reducing the price, and making the bottles 30 percent larger. He then advertised Pepsi as the Cola for the younger generation, positioning it as the "opposite" of Coke. Within a few years, he had 40 percent of the market share and Coke has never regained its overwhelming dominance.

Taking a cue from Pepsi, several decades later the Seven-Up Company embarked on a similar strategy. It positioned its soft drink as the opposite to Coke and Pepsi. Instead of being dark brown it was clear. And they called it the Uncola, a clever way of branding it the opposite of cola. Sensing another major competitor, the Coca-Cola Company purchased Seven-Up.

In the 1920s Coco Chanel changed women's fashion forever by

going to the opposite place to find ideas for a new clothing style. She borrowed from men; her famous Chanel suit—an elegant pairing of a knee-length skirt and trim but masculine jacket—created a fashion sensation that's still copied today. Over her career she appropriated fabrics, styles, and articles of clothing worn by men and combined them with feminine elements, defining a look that was both defiant and elegant at the same time.

So, take your best ideas, the best ideas of your competitors, and do the opposite. When I was a successful entrepreneur, people would ask me, "How do you come up with an idea for a new company?" I'd tell them, "Study your competitor, the most successful company in your industry, and then do the opposite." You can't beat a bigger, stronger competitor by copying them exactly, but you can often beat them by doing the opposite, because the opposite is hard to defend against. It's difficult for a market leader to copy an idea opposite to theirs. Al Ries and Jack Trout call this "finding a weakness in the leader's strength." And while the technique appears exotic, it really isn't; it's just derived from a close study of your competitor.

The opposite place doesn't always work. You won't always find a solution hidden there, but you should always explore the possibilities. Once you've gathered the opposite material, take a step toward more distant lands and begin gathering more exotic stuff.

Borrowing from a Similar Place

You should always pass through those subjects that are similar to yours on your way to subjects that are seemingly unrelated to yours. If you have a navigational problem with your tax software, then examine how Yahoo! solves this problem. If you've got a steering problem with your new automobile, then see how the motorcycle people have solved it. If you're a novelist and struggling with character development, then look at how a screenwriter solves his character development problems. Different places, but similar. You'll find that these places are usually solving comparable problems.

The Google guys borrowed ideas from library science, a subject similar to Internet search but not directly associated with it, in order to develop their algorithm for ranking Web pages. Using the bibliography concept of "citations" and comparing it to Internet links, they'd think of their search engine the way a librarian or researcher thinks of bibliographical references. The more references a book gets from other books, they reasoned, the more important the referenced book. According to Larry, a Nobel Prize–winning paper will have ten thousand other papers that mention it. Likewise, the more links a Web page gets the more important the Internet site. Larry and Sergey used their definition of the problem to find another subject by which to structure their thoughts. They established an important metaphor: Finding a page on the World Wide Web is like finding the page inside a massive library. Both the Web and the library house millions of pages. This metaphor would lead to innovations in Internet search that would create one of the most important tools on the entire World Wide Web.

Darwin, too, looked for ideas outside his own subject. As he sailed the South Pacific, he studied more than just biology. He pestered Captain FitzRoy about celestial navigation and the operation of a complex nineteenth-century sailing vessel. He studied the weather. He read all the books the captain had stowed on board. Darwin was intrigued by one book in particular that FitzRoy had acquired directly from the author before they had set sail. The book was called *Principles of Geology* and it was written by Charles Lyell. Unbeknownst to Darwin and FitzRoy, the book was creating a stir in England because of its controversial nature and the challenges it posed to the biblical understanding of creation. As the *Beagle* gently rocked back and forth, anchored off the coast of Ecuador, we can imagine young Charles enthralled by this book and the revolutionary thoughts it contained. Certainly, he and the captain had long discussions about its content. Darwin was intrigued by the similarity between his assigned profession as a biologist and that of a geologist like Lyell. They were both studying the natural sciences.

The central argument of *Principles of Geology* was that geological features such as the profile of a mountain, the shape of river, or the depth of a canyon are the result of minute changes accumulated over incredibly long periods of time. The profile of a mountain range was etched by erosion, rain, and wind's slowly chipping away at it, carrying a single grain of sand down its slope, and forming massive undulations and contours over time. The shape of a river wasn't laid by the hand of God, Lyell argued, but formed naturally, over long periods of time, by water cutting away the soil on the outer edge of a river and depositing it on the inside edge around the next bend. Over eons this creates the winding effect of most rivers. Every day the river becomes a bit more sinuous, so gradually that the change is impossible to detect on a daily or even yearly basis. And the canyon, like the mountain, is etched slowly, incredibly slowly, digging itself a little deeper with every passing day by the carrying off of individual grains of sand. The geological features people see are the accumulated effects of this gradually evolving, unnoticed process.

Lyell's theory was hard to discount because you could see the grains of sand moving down the slope or being deposited on the sandbar that forms on the inside curve of a river. However, this was radical thinking in its day. A drop of rain, surely, couldn't shape something as magnificent and strong as a mighty alpine mountain range. Most people believed that the mountains, rivers, and canyons were the result of biblical-type catastrophes like earthquakes or volcanoes or that God personally shaped the earth. In fact, most scientists at that time believed the earth was about five thousand years old, an idea borrowed from the Bible. This was not enough time for these massive accumulations to have occurred. Lyell argued that the earth was tens of millions of years old and that, given such a length of time, the mountains, rivers, and canyons have indeed evolved through these tiny little changes. Today, Charles Lyell is considered by most scholars the father of modern geology and has a prominent burial vault in Westminster Abbey. He isn't, however, considered the father of evolutionary theory, even though, in many ways, he is.

You see, Charles Darwin borrowed this radical geological idea from Charles Lyell, who would years later become his mentor and personal friend, and applied it to his thinking to create a radical biological idea. Organic material, Darwin argued, evolves just as inorganic material does: with minute changes in each descendent that, over time, accumulate to form new biological appendages like eyes, hands, or wings. He used geology as a metaphor to structure his ideas in biology. Just as Lyell pointed out that grains of sand moving down a slope would ultimately change the shape of a mountain, Darwin pointed out that small changes, mutations, could be seen in each subsequent generation of plants and animals. While inconsequential in the short term, over incredibly long periods of time these mutations combine with others to create complex organisms through evolution.

Darwin brilliantly borrowed an idea from one domain, geology, and applied it to a different domain, biology, realizing the two subjects were closely related. In retrospect, it was a rather obvious metaphor—that inorganic things evolve like organic things. What Darwin did next, though, would elevate him from the status of mere creative privateer to that of a true creative genius, and cement his place in history as one of the most effective thinkers of all time. He'd go way beyond his domain in search of another piece of information to complete the construction of his elegant theory.

Borrowing from a Distant Place

The farther away from home you travel, the farther away your subject, the more exotic the materials become. It's the contrast between the places, the distance, that make the materials exotic, and not the materials themselves. For example, to a scientist the business world is exotic because of its dissimilarity to his own. It follows, then, that the scientific world is just as exotic from the businessperson's perspective. The greater the contrast the more unusual the materials. The economist who finds the solution to his problem in a friendly game of poker is more apt to produce something of great creative value than the one who solves his problem by stealing ideas from Adam Smith or John Maynard Keynes. The businessman who recognizes his problem in the strange plants that grow on the western slope of the mountains behind his house is more apt to produce a product of great creative value than the one who solves his problem by stealing ideas from Warren Buffett or Donald Trump.

The creative privateer seeks out or stumbles upon these places in the same way that she sought out ideas from her own subject and those subjects similar to her own. She uses the problem as a map or signpost to find her way. Sometimes she consciously seeks out these places that present a problem similar to hers, especially places so far removed that the average thinker would never consider looking there for a solution. Other times, she just stumbles upon them, finding an idea when she wasn't even looking for one.

John Nash liked to play poker in his Yale dorm room. It was a way for him to relax, let his mind rest, and get away from the mathematical equations and deeply complex solutions that a doctoral candidate in macroeconomic theory works with every day. It was during one of these friendly poker games that he suddenly realized the complex problem of economics he was struggling with during the day was very similar to the problem he faced in the seemingly simple card game he was playing at night. In a poker game, each player is self–interested, but whether he wins or not, and how he plays his cards, are dependent upon how the other players play their cards. Likewise, in a capitalist economy, each player is affected by how others play their economic cards. In each case the players don't know what cards the others are holding, each has incomplete information. Nash recognized the problems were analogous. He started to study card playing more closely, borrowing poker playing strategies and brilliantly applying them to economic playing strategies, not unlike the way Darwin borrowed geological theory and brilliantly applied it to biological theory. Eventually, he'd combine this poker playing material with economic material and produce ideas that earned him a Nobel Prize in economics decades later. Most consider him a creative genius, in part, because of how he ventured so far away from his subject to solve its problems. Hell, what's farther away from Keynesian macroeconomics than the local, smoke-filled, alcohol-fueled card game in a filthy dorm room? It took a creative genius, and a man with a problem map in his psyche, to make such a connection. It took a beautiful mind.

George de Mestral liked to hike in the mountains behind his home nestled deep in the craggy Lucerne Valley of eastern Switzerland. It was a way for him to relax, to get his mind off his work as an inventor and businessman. He often had a myriad of problems randomly flashing through his mind, and these adventures into the Alps helped to erase the repetitive thoughts that occupied his consciousness. Returning home from one such journey he realized that both he and his dog were covered in burrs. As he began picking them off, he recognized a problem and solution in the seemingly benign problem of burrs attached to the fur of his Saint Bernard. Curious, he studied the burrs under a microscope and saw how the natural hooks fastened themselves to his clothing and the fur of his dog. He realized that this was a solution to a fastening problem, the same problem that a button or zipper solves. Borrowing this idea from the strange plants that grew on the western slope behind his home, he created a product he called Velcro. Most inventors consider him a creative genius, in part because of how he ventured so far away from the clothing industry to solve one of its most important problems.

While Nash and de Mestral seemed to serendipitously stumble upon solutions to borrow, others approach borrowed brilliance much more deliberately. For example, there's a formal branch of engineering called "bionics" based on the study of organic systems and how to apply them to modern technology. In other words, bioengineers purposely travel from their engineering domain to the distant domain of biology to search for ideas. A case in point is the development of SONAR and RADAR. British meteorologist Lewis Richardson was fascinated with the bats that flew outside his country home in Lincolnshire. At dusk he'd see them flying outside the house and in the fields, darting and dodging about while hunting insects for dinner. What intrigued him was that these animals were blind. Instead of using their eyes to see in the dark, they'd send out a high-pitched squeal, barely perceptible to the human ear, and then perceive objects based upon how long it took the squeal to echo off the objects. Richardson called it the ability to "echolocate" and he took this idea and borrowed it to solve the problem of locating icebergs in the northern Atlantic. Unfortunately, his patent wasn't filed until May of 1912, a month after the sinking of the *Titanic*. Following in Richardson's footsteps, others borrowed this conception over and over to solve other problems such as locating German submarines ("SONAR" is an acronym for SOund NAvigation and Ranging) and locating planes in the sky ("RADAR" is an acronym for Radio Detection And Ranging). Other bionics engineers have mimicked the lotus flower's ability to repel dirt and water to develop coatings, paints, and fabrics with equivalent properties. Engineers call these products superhydrophobic materials. You and I would call them brilliantly borrowed materials. These borrowings are so far away from their applications that they qualify as truly creative solutions.

Of course, no one has grasped the concept of borrowing brilliance more beautifully than Charles Darwin. Whether he thought about the creative process the way I think about it, as borrowing solutions from places with a similar problem, I don't know. I do know, however, that this is exactly what he did. Whether he did it consciously or subconsciously doesn't matter to you. What matters is that you learn to think the way he did, simulate the mind of Darwin, so that you can construct your own solutions to your own unique problems.

Darwin realized that tiny biological mutations could accumulate over long periods of time, an idea borrowed from Lyell. But how these mutations formed the complex biological species he had been observing was far out of his intellectual grasp. So, for the next ten years, he immersed himself in biological experiments and studies. He bred pigeons and studied the evolution of saltwater mussels. He realized that he could amplify the desirable traits of different animals through a process of selective breeding. By managing the mating process he could produce pigeons that flew faster, with larger wingspans, or differently colored feathers. This was by no means a great discovery, as people had been selectively breeding animals for centuries. What Darwin wondered, though, was how did selective breeding work when no humans were involved to manage the process? What was the natural mechanism that drove the evolutionary process?

He thought about this for years. Sometimes the problem was at the top of his mind, causing him to think long and deep about it. Other times he worked on different problems and let his central problem incubate in his subconscious. He cast about for facts, collected information, recorded alternative perspectives, and searched for new and different ideas. He read about things close to home, and he took intellectual excursions far away, all the while pondering his problems, letting them create a psychic map in his head, a map to other solutions. He read books about birds, physiognomy, epistemology, and a wide variety of other subjects. And then, in September of 1838, he had his "aha" moment when he sat down and read a popular and controversial essay by a political economist.

Thomas Malthus was a well-known, widely read British demographer and economist who had published a paper called *Essays on the Principles of Population*. In it, Malthus said that populations have a tendency to proliferate beyond their available natural resources. This is because, he explained, populations naturally grow at a geometric rate (2, 4, 8, 16, et cetera) whereas the food supply naturally grows at an arithmetic rate (2, 3, 4, 5, et cetera). Runaway population growth, he explained, was typically prevented by what he called "checks." For example, the ultimate check for human beings was starvation, but other checks included endemic disease and prolonged periods of warfare. Inherent in the Malthus hypothesis was the notion that as the population of any species grew, the members of that species would ultimately have to fight over the limited resources available.

It was at this point that Darwin screamed "aha," or the equivalent expression of a venerable English gentleman and scholar. He instantly recognized the similarity between the problem Malthus was solving and the problem he'd been turning over and over in his mind for years. This was the final piece of the puzzle he needed to complete the construction of his evolutionary theory. He now had the natural mechanism that drove the evolutionary process. It was the fight for survival that managed the evolution of species. Those descendants with the variations, the tiny mutations, that allowed them to survive and procreate would then propagate those variations, those mutations, in a natural progression of evolution. Aha, indeed.

Whether or not you believe in evolution by natural selection is not important. You don't need to believe in Darwin's theory to appreciate the magnificence of the thinking. He traveled near and far, borrowing ideas from places closely related to biology and from places so far away most wouldn't think to look there. It's a lesson in creative thinking that's not complicated. The creative thinker simply borrows solutions from those places that have a similar problem to his and then combines those solutions in a unique way to solve his unique problem. Borrow from a competitor and you're considered an intellectual pirate. Borrow from another industry and you're an intellectual privateer. But travel far away and borrow from a subject not associated with yours and you're a creative genius. Then combine all of these things in a unique way, thereby covering your tracks, and adjust them so they can best serve your purpose and solve your unique problem. It's the source of your ideas and how they're combined that will determine how others perceive the creativity of your idea.

• • • •

Sitting in my one-bedroom apartment, these are the thoughts about borrowing that I write down and begin to use as the sources of my solutions. You'll develop your own places. Remember, though, the materials you choose to solve your problem will determine the quality of the solution. Throw water on an electrical fire and don't expect to extinguish it. The material will determine the design. You can't make a fine wine out of inferior grapes.

There's a thin line between originality and plagiarism, a line often determined by the source of your borrowings. Einstein reminds us that creativity is a matter of hiding those sources. The wife of F. Scott Fitzgerald, Zelda, once said, "Mr. Fitzgerald—I believe that is how he spells his name—seems to believe that plagiarism begins at home."

For me, realizing that the true supply of creative thought comes from other thoughts was a liberating experience. It allowed me to be more deliberate in my creative pursuits. It allows me to borrow with precision and intent and to hide my sources by combining these copies with other copies and assembling them in ways no one else has ever assembled them before. First you copy. Then you create. Borrowing brilliance means building on the ideas of others, and so your creative ability is based upon your ability to define a problem and then use that problem as a map to find the ideas of others. It's from this material that you'll find a metaphor, a way to structure your new idea in terms of an existing idea. Finding and extending this metaphor will be the focus of the next step and the next chapter.

The Second Step in the Long, Strange Trip

Tom, as the head of TurboTax, hires me to work on the direct marketing program. He's under the impression I'm some kind of direct marketing expert. I don't try and convince him otherwise, even though it's far from the truth. Hell, I don't even read my own mail, let alone study it. I had developed a sophisticated direct-marketing program at Preferred Capital that Tom was familiar with, but that seemed due to pure luck, not some inherent direct-marketing skills.

Tom explains his business model to me. "We've got three sales channels," he says. "We sell in retail stores, like Staples and OfficeMax. We sell online at www.turbotax.com. And we sell directly to the customer through direct mail. The first two channels are growing. The last one's flat, not growing at all. That's yours, make it grow," he says.

I move out of my rented apartment in Arizona and into a Motel 6 in San Diego. I'm not sure if I'm going in the right direction or not. Using the thinking tools I've established, I study the problem before building a complex solution. Unfortunately, what I find troubles me. While the program isn't growing, it is nonetheless very successful and very well thought out. Using the customer list from the previous tax season (users register the product and sign up for rebates), the current program sends three mailings prior to the tax season, staggered a month apart, asking customers to order the new product and get a "free gift" in return. What bothers me is the high response rate. In direct mail, 2 or 3 percent is a successful program. TurboTax gets a much higher rate.

"You want me to improve a program that has a fifteen percent response rate?" I ask Tom.

"Yeah," he answers. Tom has just assigned me to improve the most successful direct marketing program I'd ever seen.

"Thanks a lot," I say.

Undaunted, and desperate, I continue to study the problem. Instead of trying to improve the current one, I reverse the problem and ask: *Why does it work so well in the first place? How do they produce such strong response rates?* With these questions, these problems, and others, I begin to build my overall understanding of the problem; I identify both the high-level ones and the lower-level ones.

Then I move on to the next step. I begin borrowing materials to solve the problems. I tell my family and friends what I'm doing and ask them to keep every piece of interesting junk mail they receive. After a few weeks I've got a huge pile of crap. I study it and break each one into components. I've got letters, order forms, brochures, envelopes, shampoo samples, CDs, DVDs, and hundreds of other bizarre things.

I need to come up with a good idea. It feels like my only chance to get out of my personal mess. I'm in a Motel 6 and still medicating myself with Stoli and Cran. I wonder what to do with this direct mail crap, so I ask myself: *How do I construct a creative idea out of this stuff?*

I get no answer.

CHAPTER THREE

THE THIRD STEP—COMBINING

CONNECTING EXISTING IDEAS TO FORM THE STRUCTURE OF THE NEW IDEA

Traveling back in time twenty-six years, I see myself sitting in a kitchen. I recognize it as our home on Hawthorne Drive in Sudbury, Massachusetts. Wow, I'm a good-looking kid and seem happy. Of course, I don't know I'm going to lose all my hair and gain thirty pounds in the next few decades.

The phone rings and I pick it up.

"What're you doing?" the voice says. I recognize it as my high school ally Joe Yukica.

"Talking to you," I answer.

"Well, get ready, I'm coming over to pick you up."

"Where're we going?" I ask.

"To the movies."

"I'm not in the mood," I say.

"Doesn't matter. I'll be there in twenty minutes. The movie

opened last night. I saw it with Paula. I gotta see it again. *It's the best Goddamn movie ever made.*"

It's noon when I get in the car. Joe's playing coy and won't tell me the name of the movie or why it's so important to see it on a Saturday afternoon. *Who's in it?* I ask. He doesn't know. *What's it about?* Hard to explain, he says. *Why do you want to see it again?* It's awesome, he replies. *What kind of movie?* Not sure. *Does it have a plot?* He laughs and tells me again that it's *the best Goddamn movie ever made.* His uncertainty and the wry smile on his face irritate me. It's so Joe. He likes to irritate me. It's not the first time I've heard him use the expression *the best Goddamn something* to explain things. He's got the best *Goddamn girlfriend*, drives the *best Goddamn car*, and smokes the *best Godddamn* . . . well, you get the idea.

"It's the best, Murray. *The Goddamn best!*"

"Why?" I ask.

"Just wait."

An hour later, as the theater lights dim and the movie screen comes to life, the audience begins to cheer as nine simple words appear on the screen:

A long time ago, in a Galaxy far away . . .

The only time I'd ever heard a movie audience cheer was when the shark was blown up at the end of *Jaws.* Never heard one cheer for a movie before it started, though. Obviously, Joe isn't the only one who saw it last night and I sense an excitement, an anticipation, like being at an Aerosmith concert in the Boston Garden just after they turn down the houselights but before the band takes the stage. And there is something about the words, the paradox of the introduction, that touches me. Is this sci-fi? Is this about the future? Or the past? What the hell is this? The words seem familiar, even though I've never seen them before. Joe's uncertainty, now, seems appropriate.

At about the same time I sat in this theater on the East Coast, George Lucas and his wife were on the West Coast having lunch at the Hamburger Hamlet across the street from Mann's Chinese Theater. They'd just finished a three-day marathon session editing the German version of his new movie, *Star Wars*. They were exhausted and hungry. The pressure to deliver the foreign cut had forced him to miss the premiere of his movie; in fact, it had completely slipped his mind that the picture had opened the night before. Lucas had a tendency to focus. And he hated Hollywood, the hype, the studio executives, and deal making, so he didn't care about attending his own premiere.

At some point during the meal, he looked out the window and saw Sunset Strip jammed with people and idled traffic choking the boulevard. Strange for a Saturday morning. He asked the waitress what was going on.

"It's a new movie," she said.

"A new movie?" he asked. "What movie?"

"You haven't heard?" she answered. "It's a new movie called *Star Wars*. People are coming out of it and getting back in line to see it again. I saw it last night and I'm getting off early to see it a second time, it's so cool. You guys should see it."

Lucas and his wife looked at each other in utter disbelief. His jaw dropped open and her eyes filled with tears. *You've got to be kidding me,* he must have thought. He looked out again at the turmoil raging outside, turmoil he had created. Decades had passed since Hollywood had seen anything like this. It would go on to become the most successful movie franchise of all time.

As Joe and I sat back in New England and watched, I was drawn into a world of Wookies, Jedi Knights, and Imperial Storm Troopers.

I thought to myself: *This time Joe's right, this is the best Goddamn movie ever made. It's certainly the most creative.*

• • • •

Twenty-six years later, hidden in a San Diego Motel 6, my experience in the theater with Joe pops into my head as I ponder the letters, order forms, brochures, envelopes, and shampoo samples I've collected for the TurboTax direct mail program. At first I dismiss the memory, then I realize it's another clue. Somehow it contains the answer to the question I asked myself earlier: *How do I construct a creative idea out of this stuff?* If *Star Wars* is the most creative movie I've ever seen, what makes it so? Over the past couple of decades I've seen the movie dozens of times. I have the collector's edition DVD as well as a dozen books about the making of the movie. At Preferred Capital, we would use *Star Wars* as a metaphor for our business and even made our own video version of it called *Lease Wars,* and our top sales people were called Jedi Knights. I liked to think of myself as a modern version of Han Solo, but with my hair turning prematurely gray, most of my employees insisted on calling me—much to my disliking—Yoda. *Go figure, right?* With this deep knowledge of the Lucas creation I'm in a good position to answer my own question. I know that George combined several different genres in order to create it and that he borrowed extensively from other movies and television shows. What I don't know is how he pulled all this material together and made it into a seamless, coherent creation. I ask myself more questions. How did he construct the screenplay itself? How do you make creative combinations? Is it just placing two borrowed things together?

Defining Creativity

As I sat in a business meeting at TurboTax my mind began to wander, thinking about combinations and creativity. *Could it be that simple?* I wondered. I started to analyze my ideas to see if they were merely combinations of other ideas. Each one confirmed my hypothesis. Each was the combination of something else.

Then I asked myself this question: *What's the most creative idea in the history of the human race?* As I wondered, I kept subconsciously glancing at my friend Kim Benintendo, who sat across the conference table from me. After a few minutes, she became noticeably uncomfortable and stared back at me with a *what-the-hell-are you-looking-at* expression on her face. I hadn't realized I'd been staring; I had been lost in deep thought, unconscious of my actions, busy ranking ideas, trying to identify the one that was more creative than any other. Was it Darwin's *Theory of Natural Selection?* Buddha's *Definition of Human Suffering?* Einstein's *Theory of Relativity?* Or was it *Star Wars?* Hmm, I wondered.

It was then I realized why I'd been staring at her. You see, Kim was nine months pregnant, only a week away from giving birth to her first child. She was, I recognized, carrying the answer to the question I was repeating over and over in my head. She was carrying the most creative thing in the history of the human race . . . a human being itself.

Then I asked the follow-up question: *If that's true, then is the idea of a human being a combination of two other ideas?* The answer is, of course: Yes. Human conception is the result of a mother's egg *combined* with a father's sperm to form a single cell, a new life, a new human being. In fact, most organic species are formed this way; a new organism is the combination of two existing organisms. It's the definition of life itself. With Kim's unborn child as my inspiration, I constructed an important metaphor: People are made out of other people just as ideas are made out of other ideas. This is why ideas give birth to one another and why I can say that brilliance is borrowed and always has been. After all, that's why an idea is called a conception in the first place.

Complexity as the Result of Creativity

If this hypothesis is true, if creativity is the result of a combination of existing materials, then you'd expect, over time, for ideas, concepts, things, and our entire world to become more and more complex as a result of the creative process. Of course, this is *exactly* what you find as you examine the history and evolution of ideas, concepts, things, and the world itself. It's a world that's becoming more and more complex, just as all organic species become more complex over the eons as a result of their own evolution.

According to evolutionary biologist Richard Dawkins, when living things combine, the result is a new life containing within it the genetic code of the parents. Human beings are, Dawkins tells us, replicating machines. However, humans don't make perfect copies; mistakes enter into the gene pool. First nature copies. Then nature creates. An eye, Dawkins explains, began as a simple light-sensitive nerve cell, a mutation, able to detect, very crudely, the simple patterns created by electromagnetic radiation (light). This genetic mistake proved quite effective in the fight for survival, for it allowed the organism to detect predators and mates and to hunt more effectively. Over incredibly long periods of time, billions and billions of combinations later, the eye evolved and became more complex by adding pupils, lenses, corneas, and other intricate mutations. Now shades of light could be detected, complex shapes and patterns, and two eyes combined, allowing for depth perception. Biological evolution is the result of combinations of combinations over incredibly long periods of time and the end result is immense complexity, and beauty, as is evident in the human eye and other highly evolved and highly beautiful human features.

The same is true for ideas. Intellectual evolution is the result of combinations of ideas with other ideas over long periods of time, resulting in immense complexity. When it comes to constructing ideas, human beings are still replicating machines. Remember, a Neanderthal copied the idea for the wheel from a rock rolling down a hill. Then another caveman combined the wheel with a basket and made the first wheelbarrow. Later, a Hittite soldier combined the wheelbarrow with a horse and another wheel and made the first chariot. Soon after, another person copied the chariot but added two more wheels, and a carriage was created. Centuries later, the steam engine replaced the horse, other things were added, and the first automobile was created. Every idea is an amalgam of ideas that came before, and so things become more and more complex. Today's automobile is comprised of hundreds of thousands of parts, in contrast to the original wheel and the few other parts that comprised the Hittite chariot. Today the world is far more complicated.

While Leonardo da Vinci is admired because he mastered many different subjects, the world he lived in and the subjects he studied were not as complex as they have now become. In the fifteenth century, he could read a few dozen books on mathematics and understand most of what there was to be understood. Today, it's impossible for anyone to read all the math books in the world, there are millions of them. To master a subject now a person has to specialize in a specific branch of mathematics, like advanced algorithmic analytical geometry. Of course, this leaves little time to study engineering, botany, medicine, and architecture. Sadly, there will never be another da Vinci, another Renaissance man, not because people are not as smart but because the world is just too complicated for someone to master and contribute to multiple subjects.

You can lament about this, fight it, and long for "the good old days." Or you can simply understand it, accept it, and become the creator of ideas and not just a consumer of them. You can add to the complexity and beauty of life.

Of course, complexity is the end result of an evolutionary process that is fundamentally very simple. An idea becomes complex as more things are added to it. However, like the conception of a child, the conception of a creative idea is simple at its core. It's the fusion of two existing things to make a new thing.

Making Creative Combinations

Since I've been using a construction metaphor to explain the first several steps of borrowing brilliance, let's return to it again to understand these creative combinations. As the subtitle of this chapter suggests, it's the act of combining that forms the structure of your new idea. Put simply, a "structure" is the form made by connecting one or more "components." A building is a structure that's comprised of concrete, steel, and glass. The Sears Tower in Chicago is comprised of these components, but, so, too, is the Opera House on the shores of Sydney Harbor in Australia. And yet they look completely different and serve different purposes. The difference between these structures is not in the components but in how they're connected and combined.

In order to construct new ideas, you're going to build them out of existing ideas. Like a skyscraper, an idea has components that form an overall structure. As an idea architect, you're going start a design by first borrowing a form—an overall structure—and then adjust that structure by taking it apart, rearranging the components, replacing them, or adding to and subtracting from them, and then putting them back together to form a new structure. This chapter—and so the third step—is ONLY concerned with the construction of the overall structure and NOT with rearrangement, replacement, or the addition/subtraction of components. You'll do those things in the sixth step and I'll cover them in the last chapter. For now, you're going to worry about the overall form.

Be careful not to get confused with terminology, because every entity can be perceived as either a structure or a component of a larger structure, depending upon your point of view. In other words, every building block is itself made up of other building blocks. The earth is a structure comprised of oceans, continents, plants, animals, and billions of other components. However, from a different vantage point, it's a component, just part of the larger solar system. The whole is the sum of its parts. But each part, when isolated, is itself another whole, another sum of even smaller parts. One man's structure is another man's components. And vice versa.

In business, there's confusion between the terms *strategy* and *tactics*, because strategy is structural while tactics are the components that create the strategy. From the CEO's point of view, direct marketing is a tactic that comprises his overall strategy. However, from a marketing manager's point of view, direct marketing is strategic while the envelope, list, and package are tactical. One man's tactic is another man's strategy. And vice versa.

With that said, as a creative thinker, you have a powerful intellectual tool that allows you to establish the overall form for your creative idea. In order to create, you have learn how to master the metaphor, because metaphorical thinking is how you establish the overall structure of your idea. It's with a metaphor that two ideas combine and fuse together to form a new one.

Metaphors as Structural Formations

A creative idea begins, either consciously or subconsciously, with a metaphor or analogy. By using a metaphor, a comparison of one thing to another, you intellectually connect the two things. Once this connection is made, the metaphor is extended and the two things are allowed to grow, merging the two ideas together. In the womb, a single cell from a mother is combined with a single cell from a father and life is created and grows. In the mind, a single idea from one place is combined, through comparison, with a single idea from another place. A new idea is conceived and thus grows by an extension of the comparison. This isn't a new or a radical thinking technique. It's as old as language itself. Language is conceived in metaphor, in comparing one thing, a word, with another thing, its meaning. People learn and create through metaphors and analogies. The ancient Greeks understood this. It's why Aristotle said, "It is from metaphor that we can best get hold of something fresh."

Metaphor is fundamental to how you think. It's much more than just a literary device. In the classic book *Metaphors We Live By,* authors George Lakoff and Mark Johnson say, "We have found . . . that metaphor is pervasive in everyday life, not just in language but in thought and action. Our ordinary conceptual system, in terms of which we both think and act, is fundamentally metaphorical in nature." For example, the authors explain that the metaphor "argument is war" is a common way we structure our perception of disagreements. Even if we never say "argument is war," we still imply it. We say things like: Your claims are *indefensible.* He *attacked every weak point* in my argument. His criticisms were *right on target.* He *shot down* all of my arguments. "The essence of metaphor," they say, "is understanding and experiencing one kind of thing in terms of another."

To use this tool effectively, you have to unlearn a few things before you can relearn the power of the metaphor and analogy. If you're like me, you probably think that metaphors and analogies are grammatical devices used to enrich language or prose. Metaphors, you think, are for poets, priests, and politicians, used to spice up sonnets, sermons, and speeches. A metaphor is when you imply that one thing "is" another thing. *The apple in the sky*—is a metaphor. In contrast, an analogy is when you say that one thing is "like" another thing. *The moon is like an apple in the sky*—is an analogy. In the creative world, metaphors and analogies are the same thing. They're simply describing one thing in terms of another thing. Whether you use an apple to describe the moon through either a metaphor or an analogy isn't important. What is important is that you've described one thing in terms of another, you've made an intellectual connection between two things. With metaphor, you're intellectually combining the two ideas. From this combination a new idea is structured and a new understanding of your subject is born. Every new idea is conceived this way because every new idea is a combination of existing ideas. Creative thinkers are metaphorical thinkers. Period.

Once you understand metaphors and analogies, you'll see them everywhere. You'll realize the *leg* of the chair you're sitting on was conceived by an ancient creative thinker who compared his new invention to the human body and so referred to its new components in terms of the existing components in his mind. He used the words *leg* and *arm* as a metaphor between this piece of furniture and the human form and probably used it to create the chair in the first place. Over time, with repeated use, the metaphorical quality is lost and the word or phrase assumes only a literal meaning. Today, few think of the *leg of a chair* as a metaphor; it's a literal description of a chair component. Ralph Waldo Emerson noted that the metaphor "is the fertile soil from which all language is born, and literal language is the graveyard into which all dead metaphors are put to rest." And while this is true for language it's also true for thought, knowledge, and the creation of new ideas.

Lakoff and Johnson explain that imagination is the result of new metaphorical understandings. New ways of thinking, and so new ideas, result from constructing new metaphors, combining things that have never been combined before. "Consequently," they say, "innovation and novelty are not miraculous; they do not come out of nowhere. They are built using the tools of everyday metaphorical thought, as well as other commonplace conceptual mechanisms." For example, I've structured the first part of this book with a construction metaphor, and so I've perceived an idea as a *building*. I've referred to the problem as the *foundation* of that building, existing ideas as the construction *materials*, and how the combination of these ideas forms the *structure* of your new idea. Hopefully, this helps to make the concept of creativity more coherent. Now, I could have chosen a different structural metaphor, like perceiving an idea as *food*, and this would have created a different structure for the book. I would have talked about existing ideas as *ingredients* for a new *recipe*. I would have taught you how to *bake* new ideas or how to let them *stew* for a while. Perhaps I would have taught you how to *spoon-feed* them to others in your organization. In fact, I considered this metaphor in the early stages of this book and if you look closely you can still see remnants of it. Others have perceived a creative thought as a *plant* and so taught how to *sow the seeds* for an idea. They speak of bringing an idea to *fruition* or that another one has *died on the vine*. The metaphor I choose to use, in this case a construction one, creates the overall structure for how I think about and build my ideas about ideas. It's how anyone creates anything.

Great creative thinkers master the metaphor. Unfortunately, your English teacher ruined the concept for you by relegating it to a simple writing technique. It's so much more. The story of a creative idea is the story of how its creator constructed a new metaphor as means to formulate the new idea. The more unusual the metaphor, the more unique the idea. Isaac Newton, Charles Darwin, Walt Disney, John Nash, George Lucas, and Sigmund Freud all used metaphors and analogies to construct their creations. Each thought of his subject in terms of a different subject. As you learned in the last chapter, a creative genius travels far in search of materials to construct his solutions. The genius combines this material through the use of a meaningful metaphor. Newton used an apple as a metaphor for the moon to combine terrestrial physics, represented by an apple, with celestial physics, represented by the moon, ultimately combining the works of Galileo and Kepler. Darwin used the selective breeder as a metaphor for nature to combine political economics with biology to create evolutionary theory. That's why he capitalized the word *Nature* in the *Origin of Species*. He wanted readers to perceive it as a proper pronoun, as if it were a person. John Nash used a poker game as a metaphor for microeconomics to structure a new macroeconomic theory. Walt Disney used a movie metaphor to create an amusement park. George Lucas used a mythological metaphor to structure a science fiction movie. And Sigmund Freud used so many different metaphors and analogies to construct the domain of psychology that they're now lost in Emerson's literal language graveyard. Few people think of *a stream of consciousness* as a metaphor, because it's been used so much it's perceived as a literal way to describe thought and not as a comparison to the way a river flows.

"In psychology," Freud said, "we can only describe things by the help of analogies. There is nothing peculiar in this; it is the case elsewhere as well. But we have constantly to keep changing these analogies, for none of them lasts us long enough." Freud used hydraulics, the study of fluids, to describe thought metaphorically, using terms like *flow* and *resistance*. He created a new understanding by borrowing an existing understanding of something else. When the hydraulic metaphor didn't work, when it couldn't be extended any further, he changed to a military metaphor and described thought using terms like *aggression* and *defense*. A creative thinker understands that metaphors aren't perfect and isn't afraid to abandon a metaphor when it stops working. For example, in the second half of this book I use an evolutionary metaphor and think of an idea as a *person*. Like a human, an idea *evolves over time* and ideas *give birth* to one another. The subconscious mind is like a *womb* and judgment is the *mechanism* that drives the evolution of an idea, just as the fight for survival drives organic evolution. Since creating an idea is a matter of trial and error, it was hard for me to extend the construction metaphor; instead I chose a new metaphor to explain and create this part of the process. This book is separated into two sections. The first is "The Origin of an Idea" and uses a mechanical construction metaphor to structure it. The second is "The Evolution of an Idea," which uses an organic assembly metaphor as its overall structure.

With that said, let's follow the construction of *Star Wars* to see how George Lucas structured his screenplay by masterfully combining things through the use of an overarching metaphor. He extended this metaphor throughout the movie and only periodically abandoned it when it didn't work. As the story unfolds, I'll point out how you can use the same methodology to make combinations to construct your own business ideas. After all, businessmen use metaphors to create and sell products just as scientists and writers use them to construct theories and movies. Why do you think Bill Gates calls a computer screen a *desktop*? And where do you think he got the name for his best-selling product, Windows? In business, as in all subjects, different metaphors will construct different ideas. Yahoo! used an architectural metaphor and so thought of their site as a *portal* or an *entry point* into the World Wide Web. Google, on the other hand, used a completely different metaphor to solve a similar problem. They used a library metaphor and so thought of their site the way a librarian thinks of cataloging books, using the concepts of *citations* and *page rankings* to create their company. At the end of the day, the library metaphor is more effective at solving the problem than the architectural one and is testament to the importance of choosing the right analogy.

Metaphors are like oxygen, so prevalent you don't even notice them, for most are implied, but it's this implication that forms the structure of how you think. If you master them, you master creative thinking. As in the other parts of this book, I'm merely taking things that are normally done in the shadows of the subconscious mind and bringing them into the light of consciousness so you can become more efficient at exploring them.

Constructing the Idea

The third step in *Borrowing Brilliance* is the construction step. It's making a connection between two things, combining them, and so giving birth to a new thing. In 1945, French mathematician Jacques Hadamard asked Albert Einstein to explain creative thought. Einstein paused for a moment, thought deeply, and then told him it was simply "combinatory play." So, whether you're solving a complex theoretical physics problem like Einstein, an intricate screenwriting problem like Lucas, or a simple direct mail problem like a marketing manager—it's all combinatory play. In other words, you try different metaphors to see how well they create a new combinatory structure for your idea.

The creative use of a metaphor involves three stages. First, you establish a metaphor by making a connection between two things.

You'll find those things in the materials you gathered in the previous step. For example, if you're creating a new tax software program, you'll use either TurboTax, Yahoo! or a Hollywood movie as your structural metaphor. The farther away from your subject the material you use, the more unique the metaphor, and so the more unique the solution to your problem. Second, you extend the metaphor to establish a framework for your idea and to let your creation grow. You'll see which ideas combine best to form this overall structure—which are more appropriate, in the way the library metaphor is more appropriate, once extended for finding things on the Internet, than is the architectural one. And third, you discard the metaphor when it begins extending too far and becomes meaningless. Not everything at Google is based on the library metaphor, just the overall structure. At this point, you search for new and more appropriate metaphors. Master these stages and you master the third step in *Borrowing Brilliance*.

Like any complex conception, the creative construction for *Star Wars* was an idea that took years to develop. It's an intricate synthesis of borrowed ideas, masterfully combined to create a unique structure that took movie audiences by surprise. While you and I don't have the creative talent of a man like George Lucas, that doesn't mean that we can't learn from him. You can learn to think like him; what's innate and probably taking place in the shadows of his mind, you and I will do more deliberately.

George Lucas grew up in the dusty agricultural town of Modesto, California. Far from the bright lights of Hollywood, surrounded by farms and fields, the young Lucas showed little aptitude or desire for greatness. In grade school, since his family didn't have a television, he'd show up at his neighbor's house every evening at 6:00 P.M. so that he could watch *Adventure Theater,* a television show that specialized in running old B movies from the 1930s and 1940s. In high school he cruised the streets of Modesto in his Ford Thunderbird, dreaming of being a race car driver.

George Lucas enrolled at Modesto Junior College. He and a boyhood friend, John Plummer, began to take weekend trips to San Francisco, touring the local coffeehouses, jazz clubs, and eclectic bookstores. There he discovered a completely different type of cinema, art films, or what were referred to as "shorts." As Plummer said, "That's when George really started exploring. We went to a theater on Union Street that showed art movies, we drove up to San Francisco State for a film festival, and there was an old beatnik coffeehouse in Cow Hollow with shorts that were really out there."

In these shorts Lucas found his calling. He enrolled at the University of Southern California film school, where he became, and still is, a legend. While other students struggled to produce a twenty-minute film, Lucas would turn out two or three electrifying ones complete with trailers in a single semester. He loved editing and would sit for hours running long lengths of rough film through his white-gloved hands, making cuts with a grease pencil, the scent of splicing glue dominating his small cubicle. It was said that he could make an interesting film out of the phone book.

His final project at USC was a short science fiction film called *Electronic Labyrinth: THX 1138 4EB*. It was a startling innovation for its time, a visual masterpiece, a film that seemed to have been created with a studio-size budget but had been merely bootstrapped on a student-size allowance. It won the National Student Film Festival, got the attention of the major film studios, and earned him an internship at Warner Bros. in Hollywood.

The day that George Lucas walked onto the studio lot of Warner Bros. was prophetically the same day that Jack Warner cleared out his own desk and walked off the lot into retirement. Lucas was assigned to work with a promising young director, a twenty-something-year-old guy named Francis Ford Coppola. The jovial Coppola quickly dubbed the more serious Lucas "the seventy-year-old kid" because of his solemn personality. Ironically, the childish Coppola would create serious movies like *The Godfather* and *Apocalypse Now*, while the more "mature" Lucas would direct and produce *Star Wars* and *Indiana Jones*.

Soon, Coppola and Lucas left Warner Bros. to start their own production company in northern California called American Zoetrope. The start-up capital came from Warner Bros., which was expecting five movies in return for its investment. Coppola convinced Lucas to resurrect his brilliant student film and turn it into a full-length production called *THX-1138*. Ambitious and a visual masterpiece, it's a bleak film that lacked a well-structured plot. It failed miserably at the box office. Warner Bros. demanded its money back. *THX-1138* bankrupted American Zoetrope and Coppola and Lucas went their separate ways.

Over the next couple of years, Lucas took a series of camera and editing jobs. To redeem himself, he decided to write and direct a mainstream movie. In his mind, he ventured back to Modesto and cruising the streets in his Ford Thunderbird. From these experiences, he constructed a simple screenplay with simple visuals of cars, girls, and pop music.

American Graffiti became the most successful movie of its day. Made for less than a million dollars, it grossed more than fifty million. It even garnered a few Academy Award nominations. More importantly, though, it redeemed the young director and it allowed him to return to his dream of creating an important and more complex science fiction movie.

Establishing the Metaphor

A metaphor begins with a comparison. The innovator sees one thing in terms of another thing. For you, these comparisons lie in the materials you've gathered in the second step of *Borrowing Brilliance*. Remember, since you've used the definition of your problem as a map to finding the materials, they'll already have something in common. They're already associated with each other through your perception of your problem. For example, the reason you borrowed ideas from a Hollywood movie to construct your tax software program was because you defined your problem as *creating an emotion* in the mind of the customer. So, you recognized that movies are products specifically defined to generate emotions in their patrons. From this material you established the metaphor that preparing taxes is like watching a Hollywood movie. This is in contrast to the way that TurboTax thinks about it, for they use an interview metaphor to structure their product, in other words, they think of it the way a tax accountant thinks of dealing with a new client. This new metaphor will structure a new way to think about constructing your new tax software program.

For George Lucas, his project began as a series of images running through his mind. Images of complex spacecraft, strange aliens, Apollo rocket boosters, distant galaxies, moon landings, and epic battle scenes between good guys and bad guys. It was an incoherent hodgepodge of imagery. In effect, Lucas became lost in the borrowing of pictures from B movie serials like *Flash Gordon*, in more contemporary movies like Stanley Kubrick's *2001*, and in the real-life exploits of NASA. He knew he wanted to create a science fiction movie but wasn't sure exactly what kind of science fiction movie. He wasn't sure what genre he was going to combine with science fiction to form the overall structure and plot of his movie.

Lucas wrestled with different combinations, trying to find the perfect connection between genres. He said, "It took me about three years to write the screenplay. I wrote four versions, meaning four completely different plots, before finding the one that satisfied me. It was really difficult because I didn't want *Star Wars* to be typical science fiction." Each draft was an attempt at finding a unique combination of images and a meaningful metaphor that could extend throughout the film.

One of the first attempts was combining science fiction with a spy story. In 1974, just as work on the *Star Wars* screenplay was beginning, he told an interviewer when asked about his next movie, "Well it's science fiction—*Flash Gordon* genre; *2001* meets *James Bond,* outer space and spaceships flying in it." But this was quickly abandoned as he tried to extend the Bond metaphor. The two genres didn't connect very well.

He tried combining science fiction and the American Western. But this was too much like the television show *Star Trek,* which creator Gene Roddenberry had pitched as *Buck Rogers* meets *Wagon Train.* The famous opening sequence of the TV show, "Space . . . the final frontier," is a veiled reference to this metaphor. Lucas was looking for something unique. More creative. A combination that hadn't been made before. "I wanted it to be a truly imaginative film," he said.

George Lucas was in search of the ultimate connection. He knew, both consciously and subconsciously, that the farther away the two combinations, the more contrast, the more creative and more imaginative the result. However, it seemed like it had all been done before. The first draft was about a character named Anakin Starkiller. Although the screenplay had elements of mythology, like a sect of ancient warriors called Jedi, it was far from having the well-defined mythological theme that he ultimately connected to his *Flash Gordon-*type movie. "I always have a lot trouble finding a framework with an ultrasimple base that can captivate me and captivate the public."

For the next two years he toiled, working six days a week, twelve hours a day. When asked about writing during this time, he said, "Yeah, it's terrible. It's painful. Atrocious . . ." Years later he added, "When you look at *Star Wars* it seems extremely simplistic, but it's like most successful creations: You struggle and you struggle and you struggle—for the obvious! You finally get there and you say, 'Why didn't I think of this six months ago?' But it requires quite a thought process to get down to the obvious."

It was during the third draft that he finally found the metaphorical connection he was looking for: science fiction and mythology. It was one of the images, one of the combinations, that kept popping into his mind that convinced him. He described it as "the bad guys with lasers and blasters battling the good guys with swords and stones." You see, mythology had always been there, buried deep within the material he'd borrowed, but had been obscured by other borrowings and combinations like astronaut secret agents and space cowboys. One day it just became obvious he was writing a science fiction screenplay that was a mythological fantasy. It wasn't a spy movie. It wasn't a Western. It was an ancient legend. At this point *Star Wars,* as an idea, was born. The two ideas fused together in Lucas's mind and they began to grow as one with meaning and purpose.

Mythology and science fiction was a unique and powerful combination. It was a mixture of opposites, fitting together the way a male puzzle piece interlocks with a female piece. Opposite ideas, like people, often have a strong attraction. For example, Hell's Angels, the name of the motorcycle gang, is comprised of two words in conflict with each other, opposites, and so fit together powerfully. The title of this book, *Borrowing Brilliance,* does the same thing, combining words not usually associated, and so forms a unique connection. This is why in the last chapter you consciously traveled farther and farther away from your subject to gather materials for constructing your idea. You were looking for ideas to fuse together, to mate, so that you could conceive your solution in this step. If you did your job right, you'll be able to conceive an idea comprised of opposites, ideas distant from one another, two things that interlock together and form a unique and contrasting combination. For Darwin this meant combining ideas from politics with ideas from geology. For Lucas, nothing was farther from sci-fi (stories about the future) than mythology (stories about the past).

It took two and a half years, but Lucas finally had his metaphor: an epic, mythological space opera. Now he went back to his script with fresh eyes and expanded the metaphor. He penned a new opening, a clever fusion of "Once upon a time" from Hans Christian Andersen fairy tales and the science fiction notion of a distant star system. He wrote, "A long time ago, in a Galaxy far away . . ." It was a great new beginning. Now he needed to extend his metaphor to see if it worked throughout the screenplay.

For a creative thinker, it's this initial union that conceives the creative idea. It's this connection that establishes the metaphor. Like the conception of a child, it's the simple combination of two things. But as Lucas explained, it's not always easy, and sometimes painful, to find the right combination. It took him years to see the obvious mating of sci-fi and mythology. It was right in front of him, but obscured by all of the other borrowings that floated around in his mind. The only way you know if two ideas fuse together well, or not, is to put them together and see what happens.

Ultimately, it comes back to the problem you defined in the first chapter. It was the problem that directed you to the faraway places to gather your materials. So if you stay true to this process, you're more likely to have materials that fit together well, for the problem is the glue that binds them together. It's the problem that forms the interlocking puzzle pieces. Whether you borrow an idea from your own domain or one from a faraway place, they'll be more likely to fit together because they're associated by your definition of the problem.

Of course, metaphors aren't just effective moviemaking devices, they're also effective product-making devices. For example, in 1978, Masaru Ibuka, like George Lucas, was struggling to find the right combination of components that would conceive a new idea for his fledgling company. Sony was a moderately successful Japanese firm, taking product ideas from other companies and using its low labor costs and highly disciplined manufacturing processes to create cheap "knockoff" products like televisions, home stereos, and transistor radios. Ibuka was the chairman of Sony and he longed to be creative like the American companies he was borrowing from. One of his more promising inventions was a product called the Pressman. It was a small tape recorder that journalists could take on field assignments and use as an electronic notepad. It hadn't been released to the public yet, for it had a number of technical flaws.

One day, as legend has it, he walked into a prototype lab and saw the Pressman hooked up to a pair of full-size home stereo speakers. The engineers in the lab were using it to listen to music while they worked on other projects. Ibuka asked to hear it. When it was turned on, a look of amazement washed over his face. He was taken aback by the quality of sound that came from the small device. An engineer turned it off. Ibuka shook his head. "Turn it back on," he said. He was captivated by the product. He listened for a few minutes longer, lost in deep thought, as his mind whirled, making deep connections, planting the seeds for a new idea that had something to do with the small tape recorder. Finally, he said, "Wait here," and left abruptly.

A few minutes later he came back in, out of breath, the result of running upstairs to grab a device he had seen a few hours earlier as he roamed the building. You see, in another place, on another floor, another group of engineers were working on a lightweight pair of head-phones to be used with a full-size home entertainment system. Ibuka had the small headphones in his hand. He disconnected the Press-man from the home stereo speakers and plugged it into the pair of small, lightweight headphones. He lowered his head to put the head-phones on. He switched on the Pressman. As everyone in the room watched, slowly his head began to rise, and by the time he looked up he had a huge smile on his face. The engineers were confused, but happy to see their chairman happy, and so politely smiled in return, not understanding the real source of his joy. Only Ibuka knew that he had just made an important connection, that he had just discovered something incredibly powerful. By combining two things, he had just created an entirely new product category.

The Sony Walkman, as it would be called, became one of the most successful electronic products in history. Ibuka dismissed the idea of market research and instead authorized fifty thousand units to be manufactured. This shocked his critics and stockholders, for the most successful tape recorder—ever—had sold a total of thirty thousand units over the life of the product. To everyone's surprise, except Ibuka's, the first production run of Walkmans sold out in a matter of days. Retailers couldn't keep them on the shelves. These weren't like tape recorders or transistor radios, Ibuka explained, these were "personal entertainment systems." Using a "home entertainment system" metaphor, he had combined the world of high fidelity with the world of cheap transistor radios. It was a metaphor easy for buy-ers to grasp. Sony would go on to sell thirty million of them. It caused a sensation, made Sony a household name, and changed the way peo-ple related to music. The electronic music market wouldn't see any-thing like it for decades. Not until, that is, Steve Jobs would borrow Ibuka's idea and create the iPod.

We can only imagine what was going on inside Ibuka's head as he first listened to the high fidelity sound coming from the tiny Pressman. Undoubtedly, though, it was an attribute connection that conceived the idea for the Walkman. In other words, he connected the *small* tape recorder with the *small* headphones and fused them in his mind to create a powerful combination and a powerful metaphor that was easy for him to sell. You see, similar things combine together and form meaningful conceptions just as contradictory things do because, as I explained in a previous chapter, opposites are counterparts and so fit together just as similar things do.

Extending the Metaphor

It's the extension of the metaphor that creates the overall framework for your idea. Just because two things connect together well doesn't mean that they form an effective overall structure for your creation. In order to extend the Hollywood metaphor throughout your tax software you could borrow the three-act screenplay and use it to structure the user's experience. The three acts are: I) establishing a conflict; II) escalating the conflict; and III) finally resolving the conflict. So, you begin the program with the statement "Every week the IRS has taken money from your paycheck. Now it's your turn to take it back!" This establishes the conflict, the user becomes the protagonist and the IRS the antagonist, standing in the way between the hero and her refund. The conflict is escalated by the complex tax code. The software is like a mentor to the user, teaching her how to overcome the conflict, how to interpret the tax code in the most effective manner. Income becomes a device to escalate the conflict, since it decreases the refund. Each deduction becomes a device to solve the conflict and could play like a scene in the movie. The story climaxes with a two-thousand-dollar refund (or whatever amount the user gets). As it turns out, a Hollywood movie is a pretty good metaphor for the tax software experience. I know, because I led a concept development team for a Fortune 500 company to create such a product.

However, you never know if a metaphor is going to work as a structural device until you actually extend it. It's the extension that creates a coherent structure, how everything ties together, and so leads you through the concept development process. Lakoff and Johnson say, "Thus a metaphor works when it satisfies a purpose, namely, understanding an aspect of the concept." For us, we're creating a concept for a consumer to understand, and our purpose is to solve an identified problem. If the new metaphor, once extended, does this, then we have a successful one. If it doesn't, then we look for a different metaphor.

Choosing a new metaphor, as opposed to using an existing one, is more radical thinking. It's a big change or shift in the evolution of an idea. Sometimes, it's too big a shift and you have to stick with the metaphor that your consumer already understands. This doesn't mean that you can't innovate. You can still improve the current metaphor. For example, you could keep the TurboTax interview structure, using a virtual tax accountant metaphor, but make the metaphor more deliberate. You could extend it even more. Since the TurboTax program is based on how an accountant would interview a new client, you could create a new interview for repeat customers based upon how an accountant would interview a returning client. Instead of leading her through a series of questions to find out "what she has," you could import last year's return and then ask the client, "What's changed this year?" This would make the returning-user experience much easier and more customized. I know, because I was part of a concept development team at Intuit that did just that—we called it the "Remembers Me" interview. It was merely a deeper extension of the original accountant metaphor.

Walt Disney was the master of metaphorical extension. Disneyland, for example, is an amusement park that uses a movie metaphor to structure its content. In order to extend it, he borrowed a movie production technique that he himself had created twenty years earlier. It's a technique that I've borrowed and used to create software programs and is still used in almost every movie, most television shows, and in the creation of video games. It's called "storyboarding." Walt first used it to conceive a classic cartoon called *Three Little Pigs*. A storyboard is a series of drawings that a designer uses to previsualize his creation. Once Walt had the movie metaphor, he used this storyboarding technique to create the rest of the park. He had artists draw thousands of pictures from a customer's point of view so that he could visualize the experience before he finalized any blueprints. Even the parking lot and main gates were designed with the use of storyboards. Walt wanted to visualize a family getting out of a parked car and walking toward the entrance of Disneyland just the way a director wants to visualize the opening sequence of his movie in order to perceive the emotional impact of a scene. Walt referred to the buildings as *sets* and the landscaping as *props* and employees as *cast members*.

The city planners that Walt hired to help him design the park wanted to have three separate entrances. It would allow for the more efficient flow of pedestrian traffic. No, Walt argued. He wanted just the opposite. He purposely wanted a choke point at the entrance with people bumping into each other, for he wanted this to feel like walking into a Hollywood movie premiere, like being on the red carpet. Then, he instructed his architects to design the buildings that lined Main Street, the promenade into the park, to be built using a technique borrowed from artists and set designers called "forced perspective." The first floor of these buildings is built at three-quarters scale, the second story is five-eighths scale, and the top floor is half scale. This gives the visitor the subconscious feeling of being a bit larger than normal, as a movie star feels on a movie set. You see, Walt's playing with your mind as you walk up Main Street. He's making you feel like the star in a movie but doing it in such a way that you don't even realize it. It's the masterful extension of a powerful metaphor. Walt thought that being overt about the metaphor would ruin the experience, making the fake seem fake, exposing the man behind the curtain. Instead his metaphor is extended subtly and much more effectively.

The rides at Disneyland are not boat rides and roller coasters, they're jungle cruises and bobsled plunges down icy slopes. Each ride is a movie in itself, carefully crafted through the use of storyboards. The Jungle Cruise, the most popular ride on Disneyland's opening day, is borrowed from the Oscar-winning movie *The African Queen,* starring Humphrey Bogart and Katharine Hepburn. The only difference is that it's real, and now it's a movie starring you and your family. Laid out by a set designer, it's composed of well-placed switchbacks that hide one part of the ride from another and create a cinematic-type illusion. Tomorrowland was designed with the help of science fiction writer Ray Bradbury. Space Mountain is a rocket ship ride through the darkness of outer space, not unlike an episode of *Buck Rogers* or *Flash Gordon.*

The borrowings and how they're fused together with the movie metaphor extend throughout Disneyland. Some are obvious, like the Jungle Cruise, and others are more subtle, like the forced-perspective design of Main Street. But the effectiveness and the power of the idea still permeates the park.

My daughter, Katie, and I have been to Disneyland more than two hundred and eighty-seven times (we count them) and last year we traveled to Florida to go to Walt Disney World for the first time. We were excited because we knew the Florida park was much bigger and was also a destination resort, a combination not applied to the

California park. However, we were both a little disappointed on the first day. The Florida park, while expansive, lacked the subtle feeling of the magic that exists in California.

This is undoubtedly because Walt died before ground was broken in Orlando and so wasn't in charge of the final design and construction. My first impression when I walked into the park was that something was missing. About halfway down Main Street I realized what it was: The buildings in Florida are full scale. The architects missed that subtle idea of forced perspective that Walt had used in California. They didn't understand the metaphor the way that Walt did and completely missed an important design feature. That's the difference between a great creative thinker and an average one. The great ones fully grasp the metaphor and know how to extend it throughout their creations in the most subtle and elegant ways.

No one knows this better than George Lucas. Like Disney, he is able to establish an important metaphor and then extend it both overtly and subtly throughout his creation. The creative genius extends his metaphor the way Leonardo adds depth to a fresco painting, with subtle elegance. It's the extension that creates the structure of the idea. Lucas said, "I'm not a good writer. It's very, very hard for me. I don't feel I have a natural talent for it—as opposed to camera, which I could always just do." Once he had his mythological metaphor, the writing process became a lot easier, for he'd use it to extend into a highly structured plot. He recalled a book he'd read while at Modesto Junior College about mythology. He would borrow the ideas from this book and use them to reconstruct the plot for his movie.

Joseph Campbell was an obscure college professor when in 1949 he wrote a book called *The Hero with a Thousand Faces*. In the book, Campbell argued that there's a fundamental structure to all mythological stories, a structure that all human beings relate to on a deep psychological level. This groundbreaking theory was a brilliant combination of anthropology and psychology. Campbell borrowed the work of German anthropologist Adolf Bastian, who noted that myths from around the world, whether they be religious, cultural, or simple campfire stories, all contain the same "elementary ideas." Then, Campbell used the work of famed Swiss psychiatrist Carl Jung as a metaphor to create a revolutionary way to think about mythology. Jung had proposed the idea of psychological "archetypes," the idea that there are some basic universal psychic dispositions that are common, and innate, in all human beings. Archetypes are components of a collective unconscious, a kind of shared structure to human thought and behavior. This collective unconscious, as Jung referred to it, influences things like courtship, marriage, parenting, and the rituals and beliefs about death. Campbell recognized a connection between the work of Bastian, finding commonalities in mythological stories, and the work of Jung, finding commonalities in the human psyche. Campbell began to think of myth in terms of the Jungian archetypes. Why do people who don't speak the same language, living in cultures separated by great distances and great spans of time, all enjoy the same type of stories? Why is the campfire tale told by an aboriginal elder, deep in the Australian outback, similar to the Biblical scripture repeated by the Italian bishop from the lofty pulpit of Saint Peter's Basilica? Campbell tried to answer these questions.

In *The Hero with a Thousand Faces* he posits that all myth has a common fundamental structure, which he calls the *monomyth*. Also known as *the hero's journey*, it's comprised of more than a dozen different stages, grouped into three overarching phases, and subsuming dozens of elements that are common to all stages of the myth. Common elements include things like mentors, multiple worlds, prophesies, the wearing of the enemy's skin, and animals with human attributes. The three overarching phases of the journey are the Departure, the Initiation, and the Return. Campbell was quick to point out that not every myth contains every element or every stage of the hero's journey but that most follow the three overarching phases. This structure can be found in the journeys of Jesus Christ, Prometheus, Buddha, Moses, in stories in Apache folklore, and in most Hans Christian Andersen fairy tales.

As George Lucas reread Campbell's book, he recognized the same elements in his screenplay that were contained in the hero's journey. Lucas had written about mentors, prophesies, and animals with humanlike attributes like Wookie copilots. However, what his screenplay lacked was a coherent plot. Much to his delight, Lucas realized that the plot that he needed was contained in *The Hero with a Thousand Faces*. Campbell provided Lucas with the blueprint for his movie story. So Lucas went back to work, using the monomyth as a means of gracefully extending his mythological metaphor throughout his science fiction epic. Like Disney using the forced perspective of his buildings as a psychological trick to make the visitor feel like she's in a movie, Lucas would use the Campbell monomyth as a psychological trick for making the moviegoer feel like she was watching a mythological play even as spaceships and distant galaxies floated by. It was a way for him to extend his metaphor and solve his "plot problem" at the same time.

Of course, the mythological metaphor in *Star Wars* is extended far beyond the plot and Joseph Campbell's monomyth. George Lucas brilliantly blends the images of sci-fi and mythology as well. He uses modern-looking spacesuits but accessorizes them with ancient-looking capes and robes. He takes a primeval weapon like a samurai sword and creates an elegant sci-fi weapon by combining the sword with a laser beam and calling it a "light saber." Characters have modern military titles like Commander General Tarkin and interact with others that have ancient mythological ones like Lord Darth Vader.

Some of the borrowings are subtle, like the monomyth, and others are overt, like the light saber, but together they form a highly synthesized combination.

Metaphors aren't just for movies or amusement parks. As you saw, Masaru Ibuka used the combination of a small high fidelity tape player and a small high fidelity pair of headphones to create a new product category using a "home entertainment" metaphor. He used this metaphor to market the Walkman as a "personal entertainment system," an idea that customers quickly grasped. Revolutionary new products require the same kind of metaphorical thinking that revolutionary new movies require. The automobile was originally marketed and referred to as a "horseless carriage." By marketing it this way, manufacturers made certain that people were able to understand its use and purpose. In fact, the original cars produced by men like Karl Benz and Henry Ford looked almost identical to a horse-drawn carriage, because Benz and Ford were merely combining internal combustion engines with these glorified wagons. Today, you can still see remnants of this metaphor every time someone brags about the "horsepower" of his new Porsche.

As you struggle to make your own unique combinations to solve your own unique problems, you'll want to use the metaphorical skills of men like Lucas and Disney. After all, the things they created were just products like the ones you work with. Don't forget, *Star Wars* and Disneyland are both multibillion-dollar businesses. Chris DeWolfe, one of the founders of MySpace, perceived his new creation as a network of personal Web sites using the traditional Internet concept as a metaphor for a new social networking concept. MySpace, as the name implies, is the Internet for people, not companies or organizations. It's a simple but brilliant metaphor that he was able to extend to great effect and an idea that made me say to myself, "Why didn't I think of that six months ago?" when my daughter first showed me her own Web site.

So, borrow the mind of George Lucas, for he's a master of metaphor. He knows how to conceive a new thing in terms of an existing thing and then uses this conception as the structure, or framework, for a new creation. He's also astute enough to abandon the metaphor when it doesn't work anymore, when it no longer serves its purpose, when it's no longer solving the problem identified in the first step. Extending a metaphor too far is a common mistake that rookie creative thinkers make. Discarding the metaphor is the last stage of the combinatory step.

Discarding the Metaphor

One of the first and oldest forms of surgery is a procedure known as trephining. It involves a surgical incision through the skull in an attempt to open the cranium and expose the brain. Its purpose in ancient times, it's believed, was to release evil thoughts, or spirits, from a mentally disturbed patient. Trephining is a dramatic example of a metaphor extended too far, far beyond the point at which it was solving the intended problem. Ancient surgeons perceived thoughts as being like invisible spirits and were trying to release diseased ones. It's actually not a bad metaphor and similar to the ones that Sigmund Freud and William James used. However, Freud and James understood the purpose of metaphor: that it's describing one thing in terms of another and is not to be taken literally. People tend to get carried away with metaphors and start to take them factually, in other words, fully extending them beyond their original purpose. Trephining was a metaphor gone amok. Religion, it could be argued, is another case of metaphor gone amok, its original intent lost by extending the analogies and taking them as literal and not metaphorical understandings.

As I mentioned earlier, the original automobile was perceived as a horseless carriage. Early auto engineers configured the passenger layout in the same way a carriage was laid out, with the occupants placed precariously at the front of the vehicle. This made automobile accidents mostly fatal, the extension of the metaphor carried too far and so beyond its useful purpose. Once this was understood, later designs placed the occupants more safely behind the engine instead of on top of it.

The lesson here is this: Don't confuse a tool with its function. Your objective isn't to construct metaphors, your objective is to construct meaningful and useful combinations that solve an identified problem. The metaphor is only a tool that enables you to do this. Don't forget the problem you're trying to solve and don't forget to be conscious of the problems that your new solution creates. If you're trying to solve a psychological problem with a spiritual metaphor and your solution starts killing your patients, you may want to abandon it. If you're solving an automobile problem with a wagon metaphor and your solution starts killing your customers, you may want to abandon that as well. Remember what Freud said: "We have constantly to keep changing these analogies, for none of them lasts us long enough." To master the metaphor, you have to be conscious of when it stops working. That's why, in organizing the second half of this book, I abandon the construction metaphor and use an evolutionary metaphor.

Building a world-class amusement park was much more than a matter of solving the psychological aspect of the experience. It involved a deep understanding of engineering, architecture, mass transportation, construction, landscaping, and food distribution. While Walt brilliantly extended his movie metaphor, he didn't get carried away with it either. In addition to his movie borrowings, he also borrowed from museums, parks, and city planners. His nightly fireworks show and careful attention to cleanliness were not borrowed from movies but from Tivoli Gardens in Denmark. The inspiration for the Matterhorn roller coaster wasn't from a film but from a personal trip he took to Zermatt with his wife and daughter. The steam engine that circles the park was borrowed directly from the one that circled Electric Park in Kansas City where he grew up as a child. Thousands of other things are combined to make Disneyland what it is, not just the overarching metaphor of an amusement park that is really a movie starring you and your family. If he was able to extend the metaphor, he did; if he wasn't, he didn't. Disneyland isn't "literally" a movie, it's just "like" a movie and its creator was well aware of this.

Likewise, George Lucas extended his mythological metaphor but didn't use it exclusively to construct his science fiction movie. His borrowings include much more than mythology. Darth Vader is certainly composed of mythological elements as well as science fiction ones—after all, he's half man and half machine. But he's also the combination of other things. His voice is synthesized through a standard scuba-diving regulator. His helmet is overtly stolen from the Nazi helmet worn by Hitler's SS troops. In fact, Third Reich borrowings are spread throughout the movie. The term *storm trooper* is used to describe the Imperial soldiers. And the last scene of the movie is eerily similar to the last scene in the Nazi documentary *Triumph of the Will,* produced by German filmmaker Leni Riefenstahl, which shows Hitler victoriously taking the stage at Nuremberg.

To pace the movie properly, Lucas borrowed the cliff-hanger techniques that he loved from *Adventure Theater* and the 1930s and 1940s movie serials. He conceived each act of the screenplay as an installment of one of these beloved serials by putting Luke into perilous situations that created great audience anticipation throughout the film. This had nothing to do with science fiction or mythology but was a critical element of filmmaking and storytelling, irrespective of the genre and any overarching metaphor. It was also a critical reason for the success of the movie. While Lucas considered himself a poor writer, at the same time, he considered himself an accomplished editor. This is evident throughout the film and is, perhaps, its greatest contribution to the domain of filmmaking.

Most creations are a complex combination of different things. Problems themselves are complex and are not isolated but part of a hierarchy of interconnected problems. To solve this complex hierarchy it's going to take more than just a simple metaphor, it's going to take a complex combination of various factors. However, the metaphor, if used correctly, will help you to form an overarching structure, or what George Lucas referred to as the "framework," for your solution. Mark Zuckerberg used a college yearbook as a metaphor that he extended and used to construct his new Internet site that he called Facebook. Without the metaphor your solution tends to be haphazard and not as perceptually elegant. The metaphor helps to tie all the combinations together the way that Disney tied his together with a movie metaphor and Lucas tied his together with a mythological metaphor.

Your job isn't done once you've got hold of a metaphor. The metaphor forms the overall shape of the solution, but for an idea to survive, it needs to grow, it needs to incubate and evolve into something of greater complexity and self-sufficiency. The metaphor only gives you a framework or structure but not a complete solution. That's going to take some more work.

• • • •

According to Lakoff and Johnson, "You don't have a choice as to whether to think metaphorically. Because metaphorical maps are part of our brains, we will think or speak metaphorically whether we want to or not." In other words, most of it's done in the shadows of the subconscious. But since metaphors structure how you think, once you become conscious of them, you can begin to restructure how you think by deliberately changing the metaphors you use. And that, my friends, is the definition of creative thinking.

I begin to consider combining an art form, one of the skills that separate the creative mind from the uninspired one. I realize that metaphorical thinking is the primary tool of the creative thinker. It's how combinations are conceived. I also realize that metaphors are all around me. I begin to see them everywhere, as in the software program I'm working on that uses a *key* to *unlock it*. And in the World Wide Web, which isn't literally a *spiderweb*, someone just conceived it that way. I start using metaphors to create my own ideas. Sometimes they're pretty good. Other times they're completely inappropriate or I extend them too far. Like becoming proficient at anything, it takes practice.

The Third Step in the Long, Strange Trip

Tom lets me use the office next to his at TurboTax. Once a day he stops in to see how I'm doing. On the walls I've taped up letters, order forms, brochures, and envelopes. On the desk and the credenza are shampoo samples, CDs, DVDs, and dozens of other things that I've borrowed from my friends. He looks at the stuff, shakes his head, and walks out as he mumbles something about a mad scientist.

I'm playing with components. I assemble a new idea by using an envelope from Publishers Clearing House, an order form from Columbia Music, and an audit letter from the Internal Revenue Service. That doesn't seem to work, so I rearrange it and add a brochure that I borrow from America Online. Then I alter the order form and make it an application form. It still doesn't work. This goes on for weeks. I feel like there's an idea here but I can't quite grasp it. Some combinations are truly unique but not very useful. Others are more useful, but too much like the current solution and so probably won't lead to a breakthrough idea. I'm looking for a metaphor, some overarching combination I can use to create the framework for a breakthrough direct-marketing program. I feel like it's contained in all the stuff I've borrowed, but like George Lucas, I can't see the obvious through all the clutter. Unlike Lucas, though, I don't have three years to think about this.

I'm not very good at simple math, so it takes a while for me to realize that there are millions of possible combinations of letters, offers, and packages. My head hurts from this thought. I know I can't make them all, so I do the next best thing. I decide to take off early and go to the beach.

It's sunny, summer, and San Diego. A perfect combination.

PART II

THE EVOLUTION OF A CREATIVE IDEA

CHAPTER FOUR

THE FOURTH STEP—INCUBATING

THE SUBCONSCIOUS MIND AS THE WOMB FOR A CREATIVE IDEA

Traveling back in time five years, I see myself stepping into a shower. I recognize this place as my home in Truckee, California, just over the Sierra Crest from Lake Tahoe. I'm still president of Preferred Capital. I turn on the water and wait for it to warm up. Once warm, it feels good to stand under the flow and relax. I become lost in thought, or really, lost in not thinking at all, just being, just letting the water run over my body and wash away my mind. It's empty.

And then—BANG—it hits me. I'm shot. An idea pierces my mind like a bullet through the temple. I hadn't been thinking about it, for I hadn't been thinking at all. While it kills my peace of mind, the bullet quickly transforms itself into something less invasive, then into something intriguing. It may have killed my peace of mind, but it has also given birth and left me with a new business idea.

It's simple: a zero balance bill. In other words, a monthly statement sent to a Preferred Capital customer who has an "open" line of credit with my company but isn't using it. I make money when someone finances something and I sell the paper to a bank or GE Capital. I have thousands of customers with "open" lines and I have been struggling for years to convert them to paying customers. I've tried phone calls, follow-up letters, postcards, e-mails, and faxes, but these irritated customers and made us appear desperate. Plus, it was a lot of work for little reward. For example, I would get a minor response from a first follow-up letter, then less from a second letter, and almost nothing from a third one. This was consistent with the law of diminishing returns familiar in most direct mail programs. If you mail to a list and get a 2 percent response rate, that's great, but mail it again and you'll get a 1 percent response, then a 0.5 percent reply, and so on. All my ideas to convert this valuable list were pissing these customers off.

But the zero balance bill was different. The first time I mailed it I got a 1 percent response rate. The next time I got a 2 percent response. By the end of the year, 25 percent of the list had called my company, completely breaking the law of diminishing returns. It worked better the more times I sent it. Of course, it wasn't a big surprise once you thought about it; for it was doing exactly what I wanted it to, reminding my prospect every month that he had an "open" line, but doing it in an unobtrusive way. People seemed to like bills they didn't have to pay, and it didn't look like junk mail. The zero balance bill was the final component to my Preferred Capital business model and it allowed me to double the size of my company over the next year and put millions of dollars in my pocket.

• • • •

This story comes to me five years later as I walk along the beach in San Diego, thinking about the vast number of combinations I need to solve the TurboTax direct marketing problem. It makes me ask an important question: Why do my breakthrough ideas come to me when I stop thinking about them? That sucks, because it's not predictable. It's difficult in business, or in any field, to depend on something that is seemingly so ephemeral. Businesspeople need solutions to problems and we need them immediately in a globally competitive world. So, I ask: *How do I make my breakthrough ideas appear with more consistency?* Then I ask a more specific question: *How do I recreate the shower experience, the "aha" moment, that place where the perfect idea suddenly appears?* Since I'm an amateur student of psychology, I think the answer might have something to do with my subconscious mind.

I go to Barnes & Noble and buy a dozen psychology books that deal with the subject of subconscious thought. Books by and about the major historic figures in the field such as Sigmund Freud, Carl Jung, and Viktor Frankl. I also delve into books by current thinkers like Mihaly Csikszentmihalyi at the University of Chicago.

As I read, I learn that Jung, once Freud's friend but later his rival, referred to the subconscious mind as the "shadow self," a separate personality that lurks in the darkness away from the light of conscious thought. *Cool*, I think. I also study the mind from a biological standpoint, I read books by Steven Pinker of Harvard and cognitive scientist Don Norman of Northwestern. I wonder if there is something in the construction of my mind that's reflected in the construction of my thoughts. Studying these things makes me realize that incubation is a critical part of the creative process. It's how you recreate the shower experience. It's how you use your subconscious mind as a partner in the creative process. It's about becoming a subconscious thinker.

Let me explain.

Repetitious Thought

Steven Pinker, author of the book *How the Mind Works*, says that the fundamental difference between the human brain and other brains lies in the cerebral cortex. It's unique in the animal kingdom. Your brain, he says, is actually three different brains, each the evolutionary ancestor of the other. The first brain is the reptilian brain, or the brain stem, and it controls the maintenance and life support functions like breathing, heart rate, and blood pressure. The second brain is the mammalian brain. It surrounds the brain stem and includes the limbic system, which some consider the seat of emotion and the source of the fight-or-flight instincts. The third brain is the human brain, the cerebral cortex. It surrounds both the brain stem and the limbic system and is the source of your thinking, reasoning, and consciousness. Your mind is what the third brain does.

The cortex is comprised of about a hundred billion neurons. Each neuron is a highly evolved nerve cell and it communicates with other neurons through an elaborate wiring system called synaptic connections. The neuron corresponds with other neurons by sending electrochemical signals through these connections. You experience this correspondence as a thought. So it's not the number of brain cells but how they're wired together that determines your intelligence and how you think. Each neuron is connected to tens of thousands of other neurons, making the human brain, says evolutionary biologist Richard Dawkins, the most complex thing in the universe.

At birth, Pinker explains, the cerebral cortex is mostly unwired, each brain cell independent of the others. At the same time, the brain stem and limbic system are mostly prewired and provide you with your instincts, your preprogrammed feelings and thoughts. As life begins, the cortex begins wiring itself. For example, as an infant my daughter, Katie, would listen to me as I spoke to her, and respond by calling me Daddy. As she did, she was constructing a network of interconnected cells that formed her perception of me. Later, she did the same thing as I taught her the alphabet, creating another elaborate system of cells that represented her ABCs. Every time she recited them, accompanied by a burst of electrical and chemical energy, she used the same pathway she had created earlier. Today she can recite the alphabet with little or no effort, for the network is permanently burned into her mind. This makes her a duplicating machine. She copies ideas into her brain, stored as connected nerve cells, and later uses a pathway for thought and recall. Like any path, with each use it becomes more and more pronounced. As neurologists say, neurons that fire together wire together.

Creativity expert Edward de Bono, in his book *The Mechanism of Mind,* describes thinking using a metaphor that compares the brain to a bowl of jelly. At first, the jelly is settled, its surface perfectly smooth. But when information enters, it acts like warm water on the jelly. The water forms a slight groove in the jelly and then runs off. When similar information enters it follows the path of the preformed groove. After a while, the groove becomes so pronounced that it's nearly impossible for the water to flow outside of it. These grooves are synaptic connections, and they tend to create repetitious thought because they are using the same pathways over and over. The more you repeat a thought, an idea, the deeper the groove becomes, eventually digging a canyon of thought that becomes impossible for you to think outside of.

You see, the mind is wired for repetition. You have over thirty thousand thoughts a day. That's thirty thousand bursts of electrochemical energy streaming through the various networks within your mind. How many of them are unique? Not thirty thousand, that's for sure. If you stopped and took inventory of your daily thoughts, you'd realize that you have the same few dozen thoughts over and over and over. For example, George Lucas became frustrated with the *Star*

Wars screenplay because it was difficult for him to think outside the grooves that each draft created. Once he thought of the movie a certain way he kept thinking of it that way, even though he knew it was wrong to do so. He said, "Your mind gets locked into something and it's hard to break loose, to get new ideas, a fresh point of view."

As you step through the creative process, you'll define a problem, borrow materials to solve it, and then begin making combinations with your metaphorical skills. Great, so far. But then you'll get stuck like Lucas. You'll make an interesting combination and it'll get burned into your mind. As you follow that combination, as you build it out, you'll realize that it doesn't work and that you need a new idea. However, the existing idea has formed a deep groove in your mind, an impression, that you'll keep falling into even though you know it's not the idea you want. It's the trap that eventually snares every thinker.

This is why effective creative thinking is sometimes not thinking at all. This is why you need to put an idea away. Sleep on it, as they say. For more thinking on the subject will do you no good, in fact will do you harm. You're only digging yourself deeper and deeper into a thinking hole. So, let an idea incubate. This does two things: First, it gets you out of the thinking hole. It lets the surface of your mind smooth over so you can start fresh, without the deep impressions of an ill-conceived idea. Second, it allows your subconscious mind to take over, to come in and assist you in making the combinations in an effort to solve your problem. Lucas said, "It pays to have somebody come in with fresh enthusiasm and give it a new look." For you, that someone is your subconscious mind.

The Shadow Self

The subconscious mind, according to Mihaly Csikszentmihalyi, does its thinking differently than the conscious mind. The subconscious sorts things out by making multiple combinations on different matters all at the same time. In his book *Creativity*, Csikszentmihalyi compares the subconscious mind to a parallel processing computer, where multiple operations are taking place simultaneously in different parts of the computer and then are "reconstituted" into a final solution. This allows for radically faster and more efficient computing time, and so more efficient thinking too. The conscious mind, on the other hand, thinks linearly, only able to handle one thought at a time. And this is why, in part, the subconscious is so much better at creative thinking. It can make the thousands of combinations that an idea requires for construction. It's also why Stephen King says that his ideas appear to come out of the "blue sky"; why my ideas come crashing in while I'm showering; and why the creative process is so hard to describe, learn, and teach. In ancient times new ideas were thought to be gifts from the gods; it's no wonder, the way they pop out of the shadow mind.

Sigmund Freud believed that consciousness—"awareness of thought"—was only a small part of mental life and that there was a vast dialogue taking place within the mind that never saw the light of consciousness. Hundreds, or perhaps thousands, of ideas, thoughts, and conceptions are being constructed at any one time without awareness by a person as he thinks. This is similar to Csikszentmihalyi's parallel processor, except that Freud used a different metaphor to describe the subconscious mind, since there were no computers in his day.

Freud imagined the subconscious mind as a large banquet hall, with thoughts as guests, and the conscious mind as a smaller drawing room adjacent to it. The banquet hall can hold thousands of guests, while the drawing room can only accommodate a few at a time. In the doorway to the banquet hall stands a watchman. It's his job to allow guests to enter and exit the drawing room. Once in this room, these guests, thoughts, become known to a person. However, when these guests are in the banquet hall they're on their own, free to interact among themselves. There, they're far from the light of consciousness. A person is unaware of them or what they're doing.

In the drawing room, guests are expected to act a certain way. There's a code of behavior that allows conscious thoughts to be logical, coherent, and consistent with a person's beliefs. The watchman refuses thoughts that don't adhere to this code—such as bizarre sexual fantasies— and returns them to the banquet hall or puts them away completely until they are simply forgotten. Also, the watchman turns away unrelated thoughts, like new ideas, because they're not congenial with the other guests currently in the room. According to Freud, that's why most of your thoughts are related and connected to each other. Consistency is a requirement for entry into the drawing room.

In the banquet hall the code of behavior is much more relaxed. The lights are dim and it's a party in there. Thoughts can rage, get drunk, and interact with each other without the constraints of reason, logic, or adherence to a strict belief system. Thoughts socialize, mingle, and fornicate, creating new thoughts in the process. It's like a Roman orgy. Two existing ideas meet, combine, and form a new idea, just for the hell of it. Things don't have to make sense. Beliefs are for the light of the drawing room. And the more guests the merrier. This is why the subconscious is so good at creative thought. It can make thousands of combinations below the surface, through trial and error, while life unfolds above, and conscious thought linearly lumbers along. Of course, this is why most people are uncomfortable with the subconscious mind. They can't deal with its chaotic nature, and are afraid of what may come out of it. So they place a strict watchman at the door.

Unfortunately, this kills the creative spirit. A firm watchman cuts off the primary source of inspiration. You no longer have access to the new combinations created in the banquet hall, and so the party rages on without you, or, sadly, slowly dwindles down; guests leave, are forgotten, and no longer interact, realizing that they'll never be called into the light, so what's the point? Your mind grows darker, thoughts no longer ask for entrance, and the drawing room itself dims and is left with a few, albeit consistent, but old thoughts that are repeated over and over.

Of course, you can't let this happen, not if you want to live a creative life. You have to encourage the partying, let it rage, and relax the watchman at the door so you can take a glimpse inside the banquet hall and use the children of your thoughts to solve the problems you've identified. This is a matter of establishing a working relationship with your shadow self, nourishing it, teaching it, and communicating with it. Your subconscious should be a friend, not an enemy, for without it you're left with boring, repetitive, coherent, conscious thoughts.

Three Stages of Subconscious Thought

Creative thinkers use the subconscious mind as a partner in the creative process. The clichéd image of the absentminded professor, lost deep in thought, is the image of a person in touch with the shadow places of the mind. In the fall of 1954, a young female graduate student was walking to class on the campus of Princeton University. In front of her, walking in the same direction, was a white-haired man, hunched over and lost deep in thought. Suddenly he stopped, his head snapped up, and his body became erect, as if he had been shot in the back by something. He stood there, lowered his head, and remained frozen in the middle of the sidewalk. As the student passed, the old man looked up and said, "Can I ask you a question?" Startled, and recognizing him, the young girl politely answered, "Yes, Professor." He asked, "Can you tell me which direction I was walking in before I stopped here to think for a moment?" Amused, she pointed in the direction she was walking. "Oh, good. That means I must be coming from the cafeteria and going back to my office." You see, Albert Einstein was so in touch with his subconscious mind that it would literally take over his conscious mind, erasing any meaningless thoughts or inconsequential pieces of information, such as whether he had just eaten lunch or not. Einstein could never have conceived the ideas he did without a deep, personal relationship with his shadow mind.

Of course, you and I will never achieve this kind of deep connection with the dark places of our minds. Even if we could, I wonder if we'd want to. However, as a creative thinker, I realize that using my subconscious mind not only benefits me, but that without the ability to access it, I'm like a prisoner sentenced to the deep canyons of my existing ideas, relegated to thinking the same things over and over. That's a sentence I'm not willing to accept. So I've spent a lot of time developing a personal relationship with my shadow self. It's this relationship that allows me to escape the prison of repetitive thought. It probably came naturally to thinkers like Einstein and Newton. And I'm sure George Lucas has an innate ability to reach into the depths of his mind for ideas. You and I need to be more deliberate about it.

There are three stages to subconscious thinking, to developing the shadow relationship. The first is the input stage, supplying the subconscious mind with the materials it needs and asking a series of questions. It's talking to yourself. The second is the incubation stage, letting the subconscious mind do its thing below the surface without conscious interference. It's not thinking at all. The third is the output stage, becoming conscious of a newly constructed idea. It's listening to yourself. It's about creating the right environment for listening. It's the deliberate orchestration of the elusive "aha" moment. It's digging deep into the mind, into the banquet hall, and inviting a promising new combination into the light of consciousness.

You're already doing these three things. Now you're just going to do them with purpose and meaning. You're going to be deliberate in dealing with your shadow self. While I've separated the three stages here, in reality, all three can take place, one after the other, in the wink of an eye.

The First Stage—Input

Certainly, you talk to yourself. We all have an ongoing dialogue that takes place in the confines of the mind. We don't speak out loud, that's what crazy people do. For you, it may be a constructive conversation in which you ask questions and answers magically appear. Interesting insights may come to you in the form of your own voice. Other times, it may just be an inner narrative, like a third party commenting on your life as it passes in front of you, saying things like "Isn't that sunset beautiful?" or "This guy is boring the hell out of me," or "This Starbucks employee is so slow and I'm late." For others, the narrator is a tormentor: "I'm getting fat," "I'm going to lose my job if I don't shape up," or probably something like "I'm never going to understand what this author's trying to tell me." These voices seem natural, flowing like a river, and you probably feel no particular control over them. "That's just what I think. I can't control my thoughts."

Taking control is not as hard as it may at first appear. You're not trying to control the thoughts themselves. You're not trying to control the conversation. Instead you're taking control of the subject of the conversation. In the creative process this means controlling the input, the guests that enter the banquet hall. These guests are the ones you defined in the first three steps of *Borrowing Brilliance*. From the conscious drawing room, you let in the problems you've defined, the borrowed materials you've gathered, and the metaphorical structures you've introduced into the banquet hall of subconscious thought. In other words, you feed your shadow self with problems, borrowed ideas, and promising combinations.

In essence, you're teaching your subconscious mind the creative process by consciously feeding it these things. You're showing it how to make a creative idea just as you teach it how to do other things. While you may not be aware of your shadow self, it's aware of you. It's listening, watching, and even doing things for you. If you drove to work this morning, it was your subconscious that did most of the driving, not your conscious mind. You may have been aware of the car in front of you, maybe even the route, but the gas pedal, brake, and steering wheel were controlled by your shadow self. Through repetition your subconscious learned how to apply the right amount of pressure to the gas pedal, the precise force needed on the brake, and how to spin the steering wheel to make a turn and then let it go to bring the car straight again. First, you learned these things consciously. I know, for I'm teaching my daughter, Katie, how to drive and she's still uncomfortable letting the steering wheel go, she tries to bring it back herself and so oversteers the car. Yesterday she hit the curb. She needs time to let her subconscious learn how to do these things. Many things in life are like this, controlled by the subconscious, while the conscious mind is off doing other things.

In sports, this is called creating "muscle memory." Lebron James spends hundreds of hours working on different shots: the hook shot, the layup, the dunk, and the jump shot. He repeats them over and over, taking note of his body position, his footwork, and the nuances of how he handles and releases the ball on each shot. In the game, however, he's focused on the plays as they unfold, and not his body position or how he releases the ball. Those things, to be effective, must come naturally, from his subconscious, from the muscle memory that years of repetition has created. The same is true for you as a creative thinker; you need to constantly and consciously repeat the first three steps to prepare your shadow self for the birth of a new idea. It doesn't matter how proficient you become at creativity, you'll still want to consciously feed yourself problems, borrowed ideas, and combinations, just as Lebron James still practices every day, even though he's at the top of his game.

Repeat to yourself: *1) What problem am I trying to solve? 2) What solutions can I borrow to solve it? and 3) What combinations can I make to solve it?* Write the problem out. Describe the borrowed ideas. And start making metaphorical combinations. Then periodically review what you've written. Say it out loud. Read it in the morning and in the evening before you go to bed. You're just trying to teach your subconscious the way Lebron James practices his jump shot and Katie practices her driving. Repeat it over and over to create the mental muscle memory. This is the practice; the game is creating the new idea. In business this means I keep a notebook with problems, borrowings, and promising metaphors, constantly referring back to it, knowing that I'm pushing these things deep into my mind, knowing that my subconscious is listening and that these thoughts are guests entering the banquet hall of subconscious thought. It's like planting the seeds in your mind so that they can grow, hidden from the harsh light of conscious thought, which tends to kill a young seedling of an idea.

Of course, over time, this repetition is going to gouge out a deep canyon of conscious thought and you'll run the risk of being trapped in this thinking hole. Consequently, you've got to put the ideas away. You need to leave them alone to let the surface of your mind be wiped clean, the way a Zamboni machine smooths the ice surface in between the periods of an ice hockey game. The incubation period can be as quick as a simple two-second pause in your thinking. Or it can be longer. You might sleep on it overnight, or even abandon the problem for years. It depends upon your situation and the depth of the thought canyon you have constructed.

You see, the creative thinker needs a good memory and highly disciplined mind to understand complex subjects and for gathering borrowed materials with which to construct creative ideas. And at the same time, the creative thinker needs an undisciplined and scatterbrained mind to overlook or disregard how these complex things are perceived. The ability to remember and forget: contradictory qualities and paradoxical concepts, for sure. That's why a thinker needs to incubate. To stop thinking.

The Second Stage—Incubation

By now you have a deeper understanding of, and appreciation of, the origins of a creative idea; that is, a creative idea is made up of a combination of other borrowed ideas, which, united, constitute the solution to a well-defined problem. Simple, right? Well, yes and no. Simple in concept but incredibly difficult and daunting in practice. Figuring out what to combine and how to do so can drive one insane. Four independent variables combine into millions of permutations. Since your mind is filled with millions of ideas that can potentially combine with millions of others, the possible number of permutations is incomprehensible. You can't do it consciously, not in the linear way the conscious mind thinks. You've got to use Freud's banquet hall, Csikszentmihalyi's parallel processing computer, Jung's shadow self, or Stephen King's clear blue sky.

Conscious thought and subconscious thought are not independent of each other. As I've noted, the subconscious watches the conscious mind and learns from it. In the next stage, I'll turn that around and teach you how to consciously watch the subconscious and invite its guests into the light of consciousness. In this stage, however, you need to break the relationship, sever the ties between the two minds. Incubation requires no interaction with conscious thought. Remember, you're trying to think your way out of the deep grooves, the thinking hole, that the effort of studying your subject and of developing new ideas has created. Conscious interference with the incubation of an idea will only extend that thinking hole into the subconscious mind.

You can incubate an idea in a few seconds by merely pausing your thought process, dropping its contents into the shadow mind, and then waiting for a response. It can be in the form of a brisk walk where the thinker willfully clears out her mind to rejuvenate herself. Or it can be putting the idea away and "sleeping on it." Sleep is the great rejuvenator and functions, in part, to smooth out the grooves and start the new day off with a clearer outlook, free of thinking holes. And it can be putting the idea away for days, months, and even years, allowing the shadow self to completely rethink an idea and so give birth to a completely new conception.

Conscious thinking can be a disease. Try and stop thinking. It's nearly impossible. One thought leads, or flows, into the next one, and then the next one, and so on. This constant linear bombardment, if it's allowed to go unabated, turns into a kind of mental static, all the while digging deep grooves into your psyche, grooves that lead to repetitive thought and thinking holes. As a creative thinker, you need to slow your thinking down. Regulate it. You need to add a pause in your thinking cadence. This "creative pause" is a tiny little incubation period. It's a brief moment, a second or two, in which you stop conscious thought, clear the mind, transfer your thoughts into your subconscious, and listen for a response.

Edward de Bono, in *Serious Creativity*, calls the "creative pause" a gap, or holdup, in the thinking process. The mind, he says, makes life easy by "making things routine," by forming patterns, thoughts and assessments that it can use over and over. By pausing, a person can break out of her present thinking pattern, form a new one, and so begin to construct more creative thoughts. If a stream is blocked, de Bono says, it will create a new stream, find another way to flow. The creative pause is an attempt to block your intellectual stream of thought and let your ideas flow in a different direction.

Artur Schnabel, the great concert pianist, said it best when asked what made him different from other performers: "The notes I handle no better than many pianists. But the pauses between the notes—ah, that is where the art resides!" Thinking, like playing the piano, is an art form as well. It requires a pause in between its notes; without one the music becomes static, illogical noise, no matter how well a song is constructed and performed. When one thought runs into the next one it leaves no time for reflection, no time for deviation, and no time for the subconscious mind to comment. It's just chatter.

Using this tool, pausing becomes part of your normal thinking process, a habit, not a prescribed exercise. As you contemplate, make sure a gap becomes part of your routine. How long? It depends. It can be a second or two, or it can go on for minutes. Most will be brief gaps, barely recognizable. I use this tool as I discuss ideas with friends and business partners. I will say something. Pause. And then continue. Sometimes a new thought will enter my mind and I'll get sidetracked. Although my pauses can be frustrating for friends and business partners, they're often worthwhile and lead to new thinking. Friends call me "the King of Tangents." I know it's annoying but it's an important part of my thinking routine.

Isaac Newton was known to stop in the middle of a sentence and fall into a deep trance, sometimes for hours, while friends, colleagues, and servants stood in bewilderment and thought him mad. He was unraveling the patterns of the universe and so he needed these ridiculously long pauses to restructure his thoughts, take things apart, and make new combinations that no one had ever dreamed of before. He was the least popular professor on the campus of Trinity College because he'd stop in the middle of a lecture with a creative pause that could extend for minutes, while his students sat waiting for him to come back out of the shadows of his mind. Of course, I'm not advocating such extensive pauses. In public, your pauses will last a few seconds, at best, and that's all you need. You're not unraveling the patterns of the universe.

You will, however, want to extend these pauses during your personal time, those moments you set aside for deeper reflection. You'll probably find it difficult, at first, to quiet conscious thought. Some Buddhist monks devote their entire lives to quieting the conscious mind, using techniques developed over centuries, cutting themselves off from distractions, and applying these techniques with incredible discipline. However, you can learn to slow down your thinking, and pause for a moment. This helps a lot. You need to take the "quiet mind concept" in intermediary steps. Only a fool climbs a five-thousand-foot cliff without first climbing dozens of smaller ones to develop his skills, learn to trust his equipment, and become comfortable with the exposure.

This year I climbed a route called Snake Dike on the southern face of Half Dome in Yosemite Valley that's considered one of the classic rock-climbing routes in the world. While not technically difficult, it's very exposed; you climb up a ladder of small rock outcroppings, the size of a fingernail, with thousands of feet of nothingness below you. At one point you have to smear yourself on the rock—there's not even a fingernail to stand on—and move across to a safer point. Since I'm naturally afraid of heights, the thought of doing this climb a few years ago would have made me sick to my stomach. But I did it by taking a number of intermediary steps. I started by taking rock climbing classes in Joshua Tree National Park. There I learned to smear my feet on rock, to jam a hand in a crack, and to then shift my body weight to provide the leverage to push myself up (not pull myself up with my arms). I began on small rocks, with short fifty-foot scrambles. Next I moved to "multipitch" routes, climbing up a few hundred feet. Last fall, in preparation, I went to North Conway, New Hampshire, and climbed White Horse Ledge, which has eight hundred feet of exposure at the top of the route. All of these intermediary steps prepared me physically, technically, and emotionally for the daunting task of climbing the five thousand feet of Half Dome.

Ceasing conscious thought, for me, is like trying to climb Half Dome without any climbing experience. It's not impossible, but too much to ask. My mind is infected with the disease of repetitive thinking. So I have developed a simple tool to help cure this illness. I have learned how to deliberately change the subject of my thoughts as an intermediary step toward ceasing all thought. It's easy to do. *Let me explain.*

I do this as I walk the beach or hike in the woods. Before the walk I'll choose three different subjects to think about. For example, I might choose: 1) a business problem; 2) my new girlfriend; and 3) a chapter in the book I'm writing. It works best if the subjects are not related. Then, as I hike, I deliberately choose a subject to think about, I say to myself, "Okay, let's think about Deborah." And I do. Then, after a few minutes, I'll stop walking, pause, and say to myself, "Okay, now let's think about the business problem." And I do. Then, a few minutes later, "Okay, let's think about the book."

Like smearing my foot on a rock, over and over, I'm teaching myself how to take control of the thinking process. As I said earlier, I'm not controlling the thoughts themselves, only the thinking subject; the thoughts will naturally flow from the subconscious. This is a simple exercise but a very powerful one. It helps me to get out of the thinking holes I tend to fall into. It's also an intermediary step toward the bigger goal of erasing thought, increasing the length of my creative pause, and so developing a longer incubation period.

Once I'm comfortable deliberately changing my mind, I simply replace one of my three subjects with a new one. A blank one. Nothing. As I continue my walk in the woods I've got three subjects: 1) a chapter in my book; 2) Deborah; and 3) nothing at all. Since I've already trained my mind to change subjects, it more easily falls into the blank subject, which it perceives as just another topic to think about. It's a trick, a way to train myself to create a pause in my thinking that can lead to longer pauses and ultimately the erasure of content from my mind. Sure, it's a small step, a simple one, but it works well. When you want to take the bigger steps, longer pauses, to deeply communicate with and explore your shadow world, taking these smaller steps will help; just as practicing your footwork on the small rocks in Joshua Tree prepares you for the bigger rocks in Yosemite.

This is why most creative thinkers incorporated "walking" into their daily thinking routines. They consciously or subconsciously developed thinking techniques like these. The biography of a creative thinker usually includes a story or two of walking and thinking. Thomas Jefferson, Isaac Newton, Albert Einstein, and Charles Darwin all included long walks as part of their daily schedule. If you ever get the chance to visit Darwin's country estate in Downe, England, the tour guide will show you the famous "Sand Walk" outside the house that is known as "Darwin's Thinking Path." He used the trail for his daily strolls through the English countryside in which he contemplated the human animal and his place in the evolutionary chain of all living things. Undoubtedly, he used the Sand Walk for both thinking and not thinking at all. As Friedrich Nietzsche said, "All truly great thoughts are conceived while walking." And while you and I won't take on such monumental problems, we can still use a daily walk to help us contemplate the simpler ones we have.

Once you've developed the ability to pause, become conscious of the next thought that enters your mind, the one that breaks the silence, for that's a message from your subconscious. So listen to it. Pausing is how you engage in a conversation with your subconscious mind.

A word of caution. If your subconscious is like mine, and I suspect it is, don't expect brilliance from every conversation. This is a trial-and-error process. Just because your subconscious is speaking doesn't mean it's an authority on the subject—its ideas may not be fully formed yet, they may not be ready for birth. So just listen. If they're something you can work with, great, if not, that's okay, better luck next time. Remember, it usually takes more than one encounter to get pregnant; you've got to do it over and over.

One more suggestion. If the idea that pops into your head doesn't make sense, maybe it's just a bread crumb. A clue. Follow it. Let it put you on a new train of thought. Chase the idea into the subconscious, into the shadows of the banquet hall, the way Hansel and Gretel followed their bread crumbs back home. You never know where it might lead.

In summary, the pause does two things. First, it smooths out the conscious mind, allowing your thoughts to flow in a new direction. Second, and relatedly, it allows the subconscious to speak and the conscious to listen. You can't listen if you're talking. The pause allows this communication to take place. Without it, you ramble on and on and the shadow never has a chance to get a word in edgewise.

In addition to the pause, you've got a time-honored creative incubation method that you share with every thinker in the world. It's called sleep. The cliché—*sleep on it*—became a cliché because it's extremely useful. So, I don't need to teach you how to sleep, I only need to point out its effectiveness and how to integrate it into your thinking routine.

According to Dr. Carlos Schenck, sleep expert and author of the book *Sleep*, it's a mystery why sleep is so mentally invigorating. However, he says, without it you're tired, anxious, depressed, confused, and can feel like you're intoxicated. Sleep deprivation, Schenck says, causes the brain to light up like an electrical storm with uncontrolled thought bursts, downpours of confusion and disorder. After a good night's sleep, though, the mind is clear, quiet, and still, not a ripple on the surface, like Lake Tahoe on a summer morning. It's a mind cleaned by a night's incubation period. It's a mind taken over by the subconscious. A mind prepared for creative thought like fresh ice before the start of a professional hockey game.

The awareness of a thought tends to burn the thought into the mind, digging deep grooves, the way a river digs out a canyon a little deeper with every passing day. Soon, the only way for water to flow is through the canyon. Conscious thought does the same thing because neurons that fire together wire together. However, when you sleep, these grooves are washed away, clearing the mind so it can start over with a fresh, smooth surface. That's why ideas pop into your head in the morning shower and not during an evening shower or as you drift off to sleep. Typically, mornings are better for clear thinking, for corresponding with the shadows. Evenings are for conscious thinking, for directing and repeating. So, before you go to bed, feed your mind with problems, borrowed ideas, and promising combinations for your shadow to contemplate while you sleep and dream.

Freud called dreams "the royal road to the subconscious." Both he and Jung studied dreams and considered them a meeting place between the conscious and subconscious. A dream, according to Freud, was the place where the subconscious takes the mind to the stage of awareness, where the light of consciousness is allowed to shine into the banquet hall of the subconscious. So, for you, as a creative thinker, it's a time to glimpse the shadow world. Like the thought that pops into your head after a creative pause, a dream is a message from your subconscious mind that pops in during the incubation period of sleep.

Thomas Edison used his dreams to solve problems. He'd sit in a comfortable chair, holding a heavy metal ball in his hand, place a metallic pan on the floor, and would relax, contemplate his problem, and slowly drift off to sleep. As he did, his body would loosen up, causing him to drop the metal ball into the pan. The sound of the ball hitting the pan would abruptly wake him out of his subconscious state. He was now in a half-conscious, half-subconscious state. Edison believed this to be the ideal mental state in which to solve problems.

Like Edison, throughout history creative thinkers have tapped into their dreams to solve problems. Novelist Robert Louis Stevenson's dream of being chased by the police and then turning into a monster, to pursue his attackers as an alter ego, led him to write *Dr. Jekyll and Mr. Hyde*. Friedrich Kekulé dreamed of a snake devouring itself, which led to his understanding of the benzene molecule. Albert Einstein dreamed of observing a cow touching an electric fence, which helped him conceive his special theory of relativity. And ironically, Carl Jung dreamed of meeting his deceased father, which led him to explore the idea of using dreams as a means to understanding repressed thoughts. When he was CEO of Microsoft, Bill Gates used to fall asleep during meetings and start mumbling what sounded like lines of code. It became an insider's joke at the company: that Gates even dreamed in software code, solving problems in his sleep.

With some problems, even "sleeping on it" isn't enough incubation time. The thinking holes in your mind are so deep that they're like black holes in space—the gravity of the ideas is so compelling that anytime you broach the subject you're sucked back into the hole. Imagine the intellectual struggle it must have taken for Albert Einstein, who was well versed in Newtonian physics, to break free from the pull of the great scientist's ideas and establish a new way of thinking about the physical world. It's astonishing, really. You've got the same challenge, albeit not as complex and monumental. Your industry or profession has a set of metaphors that structure the thinking for you and your associates. This thinking undoubtedly causes your mind to be sucked into the black holes of conventional thought. To break free from this gravitational force, you've got to incubate over longer periods of time. You've got to take an intellectual vacation from your problems. Leave it alone for days. Or weeks. Or even years.

Isaac Newton began thinking about gravity as a student, drawing elaborate windmill-like devices to "catch" it, perceiving it as raining down or pushing from the heavens. Then he put the subject away for years. He concentrated on mathematics and developed calculus, no small feat in itself. He also made important contributions to optics and astronomy. It wasn't until Sir Edmond Halley of the Royal Society and comet fame visited him nearly a decade after Newton's abandonment of gravity as a topic of scientific investigation and asked him to again think about the topic that new insights began to explode in his mind. The long break had allowed him to rethink the concept. Newton abandoned thoughts of gravity's pushing up in favor of its pulling down. The holes in his mind had been filled in, washed clean, by the years, and so now he was starting with a smooth surface again.

Now, if the subconscious mind is the womb in which a creative idea is conceived and allowed to grow and incubate, then it follows that, at some point, you have to give birth to the idea. At some point, the idea needs to make its way out of the shadows and into the light of day. You need to listen to your subconscious self as it tries to introduce these new conceptions to you, as it tries to give birth.

The Third Stage—Output

The creative thinker has the ability to commune with the shadow places of the mind. The shadows speak to her. Once, on a rare occasion, Isaac Newton invited two fellow professors to his chambers for wine and dinner. Newton had just been named the Lucasian

Professor at Trinity College, the equivalent of gaining tenure today. While known as an eccentric on campus, he was also well-respected and was the youngest person ever to have been named to such a prominent position at such a respected university. To show his appreciation and, perhaps, to establish some personal relationships, he asked his peers to join him for a modest celebration. During dinner, prepared and provided by servants, Newton excused himself to fetch another bottle from the wine cellar. An hour later his guests became worried because Isaac hadn't returned. According to the legend, they found him cross-legged on the wine cellar floor feverishly making notes on a piece of scrap paper. Apparently, a thought had exploded in his mind with such force that he had completely forgotten his important guests and the fact that he was hosting a party. Isaac Newton had a mythical imagination and a shadow mind that spoke to him with clarity and frequency, and with a voice so strong that it drowned out everything else.

As a creative thinker, you'll want to establish a friendly relationship with your shadow self. While you may never become as captivated by it as Isaac Newton apparently was, that's okay. You're not solving the same kind of problems. You are, however, going to nurture your relationship with your shadow self to facilitate output from the subconscious. To do this, I personally concentrate on three specific things: 1) relaxing Freud's watchman who stands at the door of conscious thought; 2) clearing out the drawing room, the conscious mind, so it can accommodate new guests; and 3) listening to the subconscious as it speaks. In other words, you need to be comfortable with new ideas, make room for them, and pay attention to them when they appear.

The watchman, Freud explained, is responsible for controlling the access to awareness. Unacceptable thoughts, such as bizarre sexual ones, may be denied access for fear they may be acted upon or interpreted as beliefs. Before he allows entrance, the watchman evaluates each guest, each idea, using criteria that include: consistency with beliefs, consistency with logic, and compatibility with the other guests in the room. He turns away not only bizarre sexual thoughts but any idea that's contrary to your belief system. The watchman denies an idea that's not logically constructed or that is not related to the other thoughts in your mind. The watchman thinks it'll confuse you. A creative thinker, on the other hand, has ideas that seem to "come out of left field" or, as Stephen King noted earlier, "Drop in from a clear blue sky." You see, the creative thinker has a completely different relationship with the watchman. It's a relaxed watchman.

To relax him, to allow ideas to freely move in and out of the conscious mind, you have to realize the difference between a thought, a belief, and reality. As simple as it sounds, in practice, it's confusing. I still have the tendency to think that my thoughts are my beliefs and that my beliefs are reality. If a thought pops into my head, I believe it to be reality. For example, if the thought *My boss is an idiot* comes to me, this means I believe "My boss is an idiot" and that in reality "He is an idiot." This is why, for most people, the ideas of the subconscious are denied access to the conscious drawing room. The watchman refuses thoughts that do not represent beliefs or what the thinker perceives as reality. This creates a quandary: Since a new idea is, by definition, a new perception of reality, it'll never be allowed access to conscious thought. How many of your great ideas has the watchman prevented from the light of conscious awareness? To become creative, you must establish new relationships with your own thoughts and the watchman that introduces them.

William James, the father of modern psychology, considered a person's relationship with his thoughts as the barometer of psychological health. He imagined this relationship as a continuum as shown in Figure 2.

FIGURE 2: THE RELATIONSHIP WITH THOUGHTS

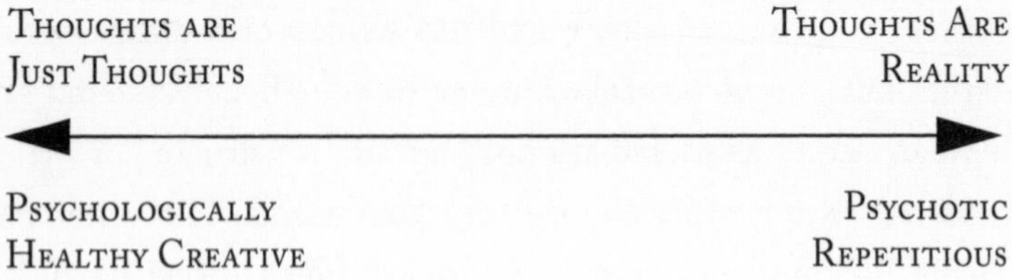

THOUGHTS ARE   THOUGHTS ARE
JUST THOUGHTS   REALITY

◄──►

PSYCHOLOGICALLY   PSYCHOTIC
HEALTHY CREATIVE   REPETITIOUS

On the far right of the continuum is the place where thoughts represent reality. Here thoughts, beliefs, and reality are all the same. Every thought represents a belief and a belief is perceived to be reality. This is the clinical definition of a psychotic. If he imagines walls are melting, then, to him, they're actually melting. As a result, a psychotic's thoughts often become repetitious, like Billy's in the Westborough State Hospital. He thinks the same thing over and over. This reinforces his reality and makes it appear more genuine and factual. A new thought, God forbid, could bring on a new and possibly terrifying reality. For the psychotic it's safer to think the same thing over and over. The watchman at the door is so concerned about the possible ramifications of admitting a new guest that he admits no guests. No new thoughts are allowed.

On the far left of the continuum, thoughts are nothing more than harmless notions. Thoughts are just thoughts. By themselves, they have no intrinsic value, unless a person accepts them as truth. William James considered this to be the epitome of psychological health. A person with this relationship to his thoughts is never tortured by them, for he realizes that his thoughts do not represent reality, that he has the right to accept or reject any idea. This attitude, of course, opens up the banquet hall of subconscious thought, for thoughts present no danger. In these circumstances the watchman vanishes. Guests in a steady stream enter and exit the drawing room, introducing themselves to the conscious light and then either moving on or staying for further consideration. In effect, you consciously assume the watchman's duties. The place has now become healthy and creative. In other words, as a creative thinker, you've established an "apathetic" relationship with ideas, not caring too deeply about any of them.

Of course, this place creates a new problem. The rush of the "aha" moment is intoxicating and makes you fall in love with your idea. The idea motivates you, forces you to think deeply, frequently, and with great care. Yet I'm advising you to establish an apathetic relationship with it. This seems counterintuitive, paradoxical. But it isn't. If you value an idea too much it becomes a belief, a reality, written in stone, and so part of the criteria by which the watchman filters and rejects ideas. People were surprised by Einstein's apathy toward his first *Theory of Relativity;* some took it as modesty, others as vanity or false modesty. However, it was just a healthy relationship he had established with his thoughts. Einstein knew there was another idea behind his first Theory. He had to let go of any belief in the first one in order to access the next one, the second Theory of Relativity. A kiss is just a kiss for lovers and a thought is just a thought for thinkers, beautiful for sure, but nothing more than what it is. Getting attached to any one thought stops others from coming to you. Let them go. It's about the journey and not the destination. You can't travel to the next idea if you want to stay and admire the last one. You've got to doubt everything, including your own ideas.

In creative thought, doubt is good. Doubt produces creative efficiency. Without it, the world is right, no need for any new thought, any new idea, and so creativity dies, aborted, before it's got a chance to be born. Socrates, Thomas Jefferson, Emily Dickinson, and Jesus Christ were all great doubters and great creative thinkers. Nietzsche said, "Great intellects are skeptical."

If doubt is good, then it follows, unwavering belief is bad. Be wary of it. The Dark Ages were times of little or no creative progress. They were also a period of great belief, a time when the church and the political system were one and the same. Belief was a requirement for membership in both. To express doubt was an act of treason, punishable by death. It was a time when learning and creating ceased. Everything one needed to know was contained in a single book—the Holy Bible. Doubt the book, lose your life. Pretty simple. Belief killed ideation.

Doubting relaxes the watchman at the door. It allows ideas to flow freely between you and the shadows. If you're a *Christian,* then read *On the Origin of Species.* If you're an *atheist,* read the *Koran.* If you're a *Republican,* listen to a *Democrat.* If you're a *Democrat,* listen to a *Republican.* See what you can learn. Expose yourself to opposing views, not to change them, but to teach yourself that it's okay to listen, that it's okay to send a subconscious thought to yourself that may conflict with your beliefs. This will facilitate the exchange of ideas within yourself. It'll help you to create and foster an open mind. Remember, Einstein doubted the Theory of Relativity. He knew it was the only way to get to the next idea, to move his thinking forward. Strong, unwavering beliefs kill creative thought. They're psychologically unhealthy and should be left to priests and politicians.

At this point, you might be asking yourself: *If that's true, then why are so many creative people mentally disturbed? Why is the clichéd image of the tortured artist, the tortured soul, on the verge of insanity, so prevalent in our culture?* Ernest Hemingway put a shotgun to his head and took his own life. Sylvia Plath and Virginia Woolf also killed themselves. Kurt Cobain and Heath Ledger are more recent examples. William Styron, Mark Twain, Edgar Allan Poe, and Robert Lowell struggled with serious bouts of debilitating depression. And Vincent van Gogh was clearly disturbed, cutting off his own ear, alienating family and friends, and ultimately taking his own life. These examples seem counter to the idea that mental health is conducive to creative thought. But it isn't.

Mental health is not necessarily conducive to creative thought; it's the relaxation of the watchman that's conducive to creative thought. Creative people who are mentally ill, like the psychologically healthy, also have a relaxed watchman; they're letting thoughts flow in and out, except they adopt these thoughts as beliefs and take them as reality, which only perpetuates the mental disorders. John Nash, the subject of the book and movie *A Beautiful Mind*, who was the victim of schizophrenia, stated it best when he said that the voice that gave him solutions to complex mathematical problems was the same voice that told him that Russian spies were following him. In other words, Nash believed that anything his subconscious told him was the truth; it was reality, no matter how bizarre or ridiculous. In fact, thinking about it this way, it's not surprising that the creative genius has a tendency toward insanity, for he begins to trust his subconscious. It has, after all, provided so many "beautiful" concepts: beautiful books, beautiful paintings, and beautiful mathematical equations. I can see why John Nash and Ernest Hemingway became tortured by the thoughts served up by their shadow selves. Hell, if my shadow had helped write *For Whom the Bell Tolls*, I might start to believe everything it had to say as well. I can see how it takes one over, especially a creatively beautiful mind.

Okay, now you've got a relaxed watchman, one that allows unfinished ideas, contrary positions, illogical conceptions, or incomplete metaphors into the drawing room. But you've still got another problem to solve before ideas start flowing more fluidly. You've got to clear out the drawing room of other thoughts. Remember, Freud pointed out that consciousness is limited, it can only handle one thought at a time. Csikszentmihalyi called it consciousness linear processing, in contrast to subconscious thinking, which he referred to as parallel processing. A new thought cannot enter the conscious mind while it's overwrought with incessant, repetitive, linear static.

You solve this problem using the thinking techniques I described in discussing the incubation stage a few pages ago. The creative pause, a simple gap in your thinking, allows your subconscious mind an opportunity to speak. Longer pauses, like those I create on my daily walks, provide even more opportunity for my shadow mind to present a new idea. I am intentionally clearing out conscious thought so that subconscious things can bubble up. In essence, I am re-creating the shower experience I described in the opening story of this chapter. This is why the second stage—incubation—leads directly into this, the third stage—the output of shadow thoughts.

Eckhart Tolle, in his popular book *The Power of Now,* advocates the cessation of incessant thinking and says it's achieved by living in the present moment. According to Tolle, conceptual thought, what we've been talking about in this book, is very limited. He describes the same relationship between the conscious and the subconscious that Freud and Csikszentmihalyi do, except he uses a third metaphor. He says that conscious thought is like the surface of a lake while subconscious thought is contained in the depths of the lake. To access the depths, a person has to let go of conscious thought and sink into the depths of subconsciousness. Living in the present moment opens up the depths to a creative thinker.

The practice of the "power of now" is simply in quieting the logical mind, silencing conscious thought. There are many ways to do this, I've touched on a few in this chapter, but anything you do that focuses you into the present helps. For example, I put myself into dangerous situations, like climbing the summit ridge of Mt. McKinley or standing on a thumbnail-size ledge three thousand feet above Yosemite Valley, not to prove myself, but because the experience itself is so vivid and powerful. My mind is focused on the next step and living in the present moment, for I realize that a lapse in concentration could mean death. Some people meditate to find the present moment. Some people use sex as a path to the present, for few minds wander in the throes of an orgasm. Others find it on a sandy path in the forest behind their home or in the silence of a cathedral in the presence of their God. Not "thinking" and just "living" is, according to Eckhart, the ultimate human experience and is a means for you to get in touch with the shadow self and the immense lake of intelligence hidden beneath its surface.

This leads to the third and final aspect of creative output. It's simple: Listen to your mind when it hands you an idea. Sometimes it's obvious, you go "aha" when the idea pops into your head, like when you're taking a shower. The moments following a pause in your thinking are the moments when your subconscious talks, and so you should be listening. Other times the idea needs to be coaxed out. You can feel it before you perceive it. So, you have to learn to recognize these emotions as the language of the subconscious. They're ideas trying to break into the light of consciousness. And finally, you have to be aware of new ideas that surface disguised as other things, like misunderstandings that are really subconscious communications.

According to the late biologist Stephen Jay Gould of Harvard University, the cerebral cortex, the thinking brain, evolved out of the mammalian brain, the limbic system, the emotional brain. They're wired together, and so work closely with one another. Thoughts create emotions and emotions can create thought. It works both ways. You know that. As a creative thinker, you need to become aware of emotions, of how you feel as you think, for emotions can be the language of the subconscious mind. Have you ever felt depressed even when upbeat thoughts are dominating your thinking? Felt down for no particular reason? Well, in many cases, it's the thoughts in your subconscious mind that are making you feel that way. The subconscious mind uses the limbic system, the emotional brain, to help surface its thoughts into the conscious mind. In this case, your unexplained depression is a negative subconscious thought that hasn't made the light of consciousness.

In Malcolm Gladwell's book *Outliers* he interviews Chris Langan, a man who has an IQ of 195, more than thirty points higher than Einstein's. Langan explains his creative process to Gladwell saying, "Sometimes I realize what the answer is because I dreamed the answer and I can remember it. Other times I just feel the answer, and I start typing and the answer emerges onto the page." In other words, Langan feels the idea that's in his subconscious mind before he's able to consciously perceive it. Emotion is like a fishing line used to catch the ideas deep below and bring them up. If you break that attachment, you've just severed the fishing line and are now unable to capture these ideas. Likewise, if you ignore your emotions in the creative process, you're ignoring the fishing line and the potential idea that you've hooked below the surface.

So, as a creative thinker, use your emotions. Get excited. Get angry. Get happy. Like Langan, I often feel an idea before I get it. I feel something brewing inside and it motivates me, it keeps me going, exploring, connecting, combining, structuring and restructuring. While some ideas come like being shot in the head, others come like the tides; washing in and slowly drowning me. These waves begin with emotion, not with a conscious thought projectile. Be aware of them. Emotions, like dreams, are just another royal road into the shadows of the subconscious. By all accounts, Steve Jobs, Isaac Newton, Walt Disney, Marie Curie, and Bill Gates used emotions. They thought *and* felt. So, too, should you. Watch the bobber in your mind and be ready to reel in an idea when you hook it deep below the surface. That hook and line are the emotions you feel in the creative process.

The last road into the shadows isn't another royal road, it's more like a secluded back street or hidden pathway. Freud was fascinated with these hidden corridors and called them "parapraxes." Today, you and I refer to them as "Freudian slips." They're misplaced words or accidental remarks, like calling your wife Camille when her name is Terri. Or referring to the hard drive on your computer as the hard . . . well, you get the idea. He considered these places were subconscious thoughts leaked into the state of consciousness. The place where a thought has escaped, cleverly dodging the watchman at the door. For us, they're pathways into the subconscious mind that can lead to new ideas.

In the creative world, there are three different types of parapraxes, or misunderstandings, that can help the creative thinker. There's the classical Freudian slip of the tongue—you say something you didn't mean to say, but it's actually a good idea. Then, there's the Freudian slip of the ear—you hear something incorrectly, but it's actually a good idea the way you mistakenly heard it. And finally, there's the Freudian slip of the eye—you read something incorrectly, but it's actually a good idea the way you read it. I've experienced all three and so have become sensitive to them. For example, a graphic designer asked me why my software product flowed "down." It didn't, but it was an interesting idea, one that led to a major innovation. Later, I found out I'd misunderstood him. He had asked me if a link opened down, not if the product had flowed down. The mix-up, I believe, was really my subconscious feeding me a new idea, a Freudian slip of the ear.

Alexander Graham Bell's invention of the telephone was due, in part, to a Freudian slip of the eye, a literal parapraxis. He read a book by German scientist Hermann von Helmholtz, called *On the Sensations of Tone*, which said vowel sounds could be reproduced using electrical tuning forks and resonators. However, Bell, who couldn't read German well, misinterpreted the passage and thought it said vowel sounds could be reproduced using electrical wire (not electrical tuning forks). This misunderstanding made him try to re-create speech using electrical wires, which he ultimately did. According to Bell, "If I had been able to read German, I might never have begun my experiments in electricity." *Was it a misunderstanding or Bell's subconscious mind leaking into the conscious world that created the telephone?* You'll never know. But you should know that misunderstandings, spoken ones, audible ones, and literal ones, are often the voice of your subconscious mind. When you spot one, ask, *Does this misunderstanding really contain an idea?* You'll be surprised how often it does. I sometimes wonder how many ideas I've missed by disregarding these parapraxes as just simple misunderstandings. Now, before I let any misunderstanding pass by me, I stop and examine them as possible notes from my subconscious mind.

As you become more aware of your thoughts and more aware of the presence of the other "you," work to establish a healthy relationship, just as you would do with any other being. A word of caution: Your thoughts are not you. You are the producer of the thoughts. You work to establish a relationship with the producer of the thoughts, not with the thoughts themselves. A kiss is just a kiss. You establish a relationship with the kisser, not the kiss itself.

For example, this is the bond I establish with myself: I listen to my subconscious and take what it has to say very seriously. At the same time, I understand it works through trial and error, that there are millions of possible combinations of ideas, and it's serving me one for consideration. I don't treat every thought as a Newtonian insight, I know that Newton didn't, but I take the time to listen, and I respect what it has to say. More often than not, I thank it for the input, and then ask it to think some more on the subject. So it does. It goes back to work and isn't afraid to offer another idea, even if it sucks. We both know that's just part of the process.

This is like the relationship I have with a friend or my daughter. It's called mutual respect. I listen. I consider. And I take what works for me. My daughter listens. She considers. And she takes what works for her. The point isn't to establish this kind of relationship, the point is to establish your own personal relationship, however you define a healthy one, with your shadow self so you can begin to communicate with it more effectively. Establish the bond so you can work together with your subconscious mind to solve your problems by combining borrowed ideas from places near and far. It's very good at doing this.

• • • •

I have no formal training in psychology, but I've become fascinated with the subconscious and work to establish a deeper relationship with mine. It's the mother of my ideas, for it gives birth to them. I realize I've got to teach it how to construct ideas: to solve problems, borrow materials, and make creative combinations. I realize it needs time to think, to incubate. I recognize the watchman in my mind, that he's screening ideas, so I work to make his duties more conscious, to stop the blind screening process, to let me do it consciously. Finally, I know I've got to stop thinking, stop talking to myself, so that my subconscious has a moment to speak. I start going for long walks. Then longer ones. I take to the local mountains east of San Diego for day-long hikes. It allows me to stop thinking, clear out my mind, and make room for new ideas.

The Fourth Step in the Long Trip

As I climb the eastern slope of Mount San Jacinto, I'm momentarily stunned, something's wrong. I stop in my tracks. I realize that I've been shot in the back of the head by another idea: an idea for the TurboTax direct marketing program.

"Send them the software, you idiot!" the bullet screams at me.

"What?" I ask.

"Why are you sending an order form? Why are you asking for the sale? Just send the software and get it over with, for Christ's sake!" my shadow says. "They're going to buy it anyway."

Such were the conversations in my mind that day that led to the TurboTax MyCD program. Tom had asked me to "make it grow," and this is how I did it. You see, the old TurboTax direct marketing program had such a high response rate (over 15 percent) because it was more convenient for some repeat customers to buy it through the mail than it was to go to OfficeMax or Staples. So, I kept asking myself, how can I make it EVEN MORE convenient? The answer was the bullet that hit me on the slope of San Jacinto. Instead of sending an order form in the mail, I decided to send the software on a CD-ROM and ask them to unlock it. That'd be a hell of a lot more convenient than having to process their order and then ship a CD-ROM anyway. My mind had constructed this solution out of the pieces of borrowed ideas that were sitting on my desk back at the Intuit office. For on my desk was a direct mail piece from America Online that was a CD for loading the AOL access software. It was one of the most promising ideas I'd borrowed, I just wasn't sure how to apply it to the TurboTax problem. Until, that is, I was shot in the head by it.

I start to explain it to Tom. After ten seconds he says, "I got it." I continue to tell him about it and he says, "I got it again."

So, I constructed a powerful metaphor for this new program. I conceived it as a *preorder* program. In other words, I designed it so that it looked like the customer had ordered the product from Intuit. It was a kind of assumption sale. So, with this metaphor, the designs were specifically un*marketing*-like. No *buy now* messaging. No *benefit* messaging. Just a simple, low-key introduction to this year's new product, like it was an upgrade they had already ordered.

That year we send millions of CDs in the mail. Instead of getting a 15 percent response rate like the old program, over 60 percent of the people respond. It's a huge, resounding success. Big revenues. Big profits. And big retention rates.

Over the next year, I talk to hundreds of TurboTax customers who tell me, "Oh, I just have Intuit send me the new program every year." When they say this, I know the metaphor is working, that I have *Jedi-mind-tricked* them into thinking they have ordered it when they had not. It's a hard-sell program packaged in a soft selling metaphor.

The idea is worth tens of millions to Intuit. Simple, yes. Obvious, maybe. But more importantly, it's the idea I need to get back on my feet. As my confidence goes up, my drinking goes down. Intuit gives me a nice bonus for my work and Tom says he wants more ideas, and so offers me a permanent job at the Fortune 500 Company.

"I'm not really a corporate kind of guy," I tell him, not being able to picture myself doing a nine-to-five type of thing. Besides, I've taken on some more consulting work, and now I've got income from three major companies: Intuit, Ingram Micro, and Insight. I'm making good money again.

"Well, I'm not offering a corporate kind of job. I've got something different in mind for you," he says.

"Tell me more," I say.

CHAPTER FIVE

THE FIFTH STEP—JUDGING

JUDGMENT AS THE DRIVING MECHANISM IN THE EVOLUTION OF AN IDEA

Traveling back in time sixteen years, I see myself sitting in front of a black-and-white computer screen. I recognize this place as the Structures Laboratory in Huntington Beach, California. It's part of the McDonnell Douglas Astronautics Company campus. In the laboratory are pieces of spacecraft, satellite junk, and a ballistic missile called the MX casually propped up in a corner. It's nuclear-tipped and it's scaring the shit out of the Soviet Union. It kind of gives me the creeps too.

"Colonel, I think we've got a problem," I say.

He leans over my shoulder to look at my screen. We're testing the structural cradle that holds the payload assist module (PAM), a rocket engine used to launch a satellite from the bay of the space shuttle. I'm the lead engineer on the PAM testing program. It's due to fly in a month and I'm simulating liftoff loads in this laboratory by applying pressure to it using huge pneumatic actuators. We're making sure it's safe for launch.

But now there's a problem. The stress-strain curve is starting to arc. So I halt the test for a moment. Stress is the force applied to a structure. Strain is the amount the structure expands or contracts while under this load. For ductile materials, strain is directly proportion to stress: Push harder and it'll compress more. That is, until just before it's about to break. Then it will compress in greater proportion to the amount of stress it's under. In other words, a little more stress causes a lot more strain. A stress-strain curve begins as a straight line, stress proportional to strain, but when the material begins to break strain becomes greater than stress, so the curve begins to arc. This is what's happening; the curve is beginning to bow, which means the PAM is just about to break.

Now, the PAM isn't a throwaway piece of spacecraft, it's to fly on the shuttle in a few weeks. Breaking it would mean scrapping the mission, and having to answer to NASA and the air force and make a telephone call to the White House. This is being touted as the shuttle's first "practical mission." President Reagan wants something for all the money he's spent.

"Something's wrong," the colonel says. "It should take a lot more load than that." I hadn't reached half the expected liftoff load. I check the different plots on my computer. There are hundreds of strain gauges in various places. I realize the weakness is isolated to one particular strut—the other curves are straight and unfaltering.

What I'd learn later was that, due to a bureaucratic error, this strut had not been "heat-treated." You see, the strut was made of an advanced aluminum alloy that was supposed to be "tempered," that is, exposed to a process of heating and cooling the metal that increases its strength and rigidity. Evidently, someone had forgotten to do this. Thankfully, the test identified the problem. We took the cradle apart, tempered the strut, put it back together, and successfully tested and certified it for launch.

A month later, as I watched *Columbia* lift off from Cape Kennedy, I had a wry smile on my face. Not only because I felt I had played an important role in what would become a successful mission, but because, when no one was looking, I had written my name on the underside of the PAM with a black Sharpie. I suppose that my space graffiti is still floating twenty-two thousand miles above me, sixteen years later, and will remain there long after I'm gone.

Go figure, right?

• • • •

Using the bonus from Intuit and money I collect from other consulting work, I buy a new home in San Diego County. I move out of the Motel 6 and into a nice place only a few blocks from the ocean. I'm trying to conceive a new start-up company, but my ideas seem frivolous. I've got a lot of them but no winners. Nothing I could pitch to a venture capital company. I ask myself: *Why are these ideas so useless?*

As I ponder these things, the story of the payload assist module pops into my mind. At first I'm not sure why. I think about it and realize it's another clue from my shadow self. It occurs to me that all my ideas are just beginnings, they need time to form, time to evolve. Ideas need to be tested, the way I tested a spacecraft before launching it. I've got to identify the weak point in my idea the way an applied force identifies the weak point in a structure. In the creative process, I realize, that force is judgment.

Perfect ideas, unfortunately, don't just appear fully formed and ready for implementation. The "aha" moment is the missing piece to a puzzle; not all the pieces are put together and ready to face the world. Ideas take time. They take work. And judgment is the thing that manages that work. Without it you have one shot at a solution. With it, you've got a machine gun with which you can pepper your problem with ideas, using each miss as an opportunity to reset the sights of your gun. Each shot, if done correctly, gets a little bit closer to the bull's-eye. Without judgment, though, you're blind and you have no idea how close you came to the target. Judgment, I come to realize, is what drives the evolution of innovation. So, I read books about evolutionary biology to help me better understand how it works.

Memes and the Evolution of an Idea

In his book *The Selfish Gene*, Richard Dawkins coined the term *meme* to describe a piece of cultural information, like a custom or an idea, that's passed on from person to person. He used it as a metaphor to describe how a gene evolves and is passed on, replicated, through organic generations. He derived the word from the Greek term *mimema*, which means to imitate or mimic. Over time, Dawkins explained, these memes are combined with others and form new ideas the way a gene is combined with other genes to form a new one. What he's really describing is the evolution of an idea, any idea, whether it's a theory, a story, a melody, or a product. And ideas, like genes, fight for survival. In fact, every idea fights with every other idea because there's a limited resource, only a certain number of ideas that can be passed along. Some survive and others don't.

What makes a survivable idea? Why do some ideas endure and others go extinct? In biological evolution, it's the fight for limited resources that determines survivability. In business, survivability is based on the importance of the problem being solved and how well the idea solves that problem relative to the other solutions. Solve a trivial problem and your idea will be trivial and die. Solve an important problem less effectively than another solution and your solution will die too. This is why, in the first chapter, you spent so much time learning how to understand the problem. If you solve an important problem that no one else is solving, you're guaranteed survivability.

The iPod survives and the idea is passed on from person to person because it solves an important problem—entertainment—and it does it better than the other solution (better than the Sony Walkman). It replaced the Walkman because you didn't have to buy and store the clumsy CDs—instead you could carry your entire music library with you wherever you went. It wasn't Apple advertising that made it so successful. It was people telling each other how cool it was, passing the iPod meme to millions. Every good marketer knows that word-of-mouth is how products really become successful, just as every prophet knows that's how a religion becomes successful. Memes are like genes in the creative process, ideas that evolve over time and survive by being more fit than those they compete with.

The Sony Walkman, while a great idea in its time, is just a single thing in a long evolutionary chain of things. The iPod is the grandchild of the Walkman. Like the evolution of an organism, ideas evolve from one species of products to the next through the combination of new elements replacing or being added to old elements. And like species, the descendents sometimes bear little resemblance to their ancestors. For example, it's believed that birds are direct descendents of theropod dinosaurs. Likewise, the iPod is a descendant of the ancient gramophone. Products and ideas are the descendents of the things that came before, constructed by combining in an effort to solve a well-defined problem. The automobile is a descendant of the horse-drawn carriage. The carriage a child of the chariot. And the chariot the son of the wheelbarrow. Each idea being born of the one before it, combined with new things, and incrementally improved, by further combination, as it evolves.

A creative thinker understands the evolutionary chain and that ideas just don't drop in from "a clear blue sky." Today, an idea is falsely thought of as a stand-alone conception constructed by an independent creator. This fallacy is perpetuated by modern "ownership" concepts of patents, copyrights, and trademarks. As art historian Lisa Pon has pointed out, during the Renaissance, artists freely copied the works of others but were expected to "improve" the copy as it evolved. It wasn't until the free market was established that "originals" became prized possessions and so of great financial value. Following this era, concepts of intellectual property were introduced into the creative world. Unfortunately, the result was a blurring of the true nature of the creative process. Now, because of legal restraints, thinkers are erroneously told not to copy, when copying is at the core of the creative process. This fallacious reasoning makes creativity seem like a magical thing that drops in from "a clear blue sky" when it's not. It's an evolutionary thing that makes use of previous ideas to give birth to new ones. And contrary to what you've heard, it's judgment that drives the evolutionary process, for it's judgment that determines survivability.

So, to generate ideas that'll survive, become an expert at passing judgment. Test ideas before allowing them out into the world the way I tested the payload assist module before it was allowed to be launched from the space shuttle. Without testing there is disaster, without judgment there are only ideas intent on extinction, frivolous things. The expertise you develop will be based upon your ability to adopt various viewpoints, to your being able to alter your view, apply different loads, and so determine strengths and weaknesses.

Let me explain.

Judgment as a Result of Viewpoint

In college, I borrowed my mother's new Ford Mustang for the weekend. On a whim, I drove from Sudbury, Massachusetts, to Williams, Arizona, to see the Grand Canyon. You know, it's the biggest hole in the planet's surface and I've returned almost every year since for my annual trek from the rim to the bottom and back in a single day-long push. Standing on top, I look down into a gaping hole, but as I descend, the canyon walls slowly change. By the time I reach the Colorado River, at the bottom, they've become a huge mountain range, not walls defining a hole in the ground. I'm in a valley, not at the bottom of a pit. My point of view has changed my perception of things.

As I write this book I use different literary viewpoints to explain my ideas. I begin each chapter using a *first-person* point of view by telling a personal story, using the pronoun *I* to let you know this happened to me, David Murray. Other times, I use a *second-person* point of view, using the pronoun *you* to illustrate the shift in perspective. And finally, I use a *third-person* point of view to tell a story about Isaac Newton or Marie Curie—using pronouns like *he* or *she* to employ this viewpoint. Most novels are written in either the *first* or *third* person, rarely the *second*—it's hard to narrate from the reader's point of view, since the reader knows the events described didn't happen to him. Each point of view provides a different type of story.

Even the passing of time can change how things are perceived. The most unpopular president in the history of the United States was not George Bush Jr. Not even close. Abraham Lincoln has the honor. In fact, he received less than forty percent of the popular vote and didn't receive a single vote in nine different states! Not one. He was so unpopular that nearly half of the states seceded. I don't recall any states being so fed up with Bush that they wanted out of the Union. Today, however, Abraham Lincoln is considered one of the greatest presidents in American history. Of course, with time we have a different perspective: that slavery, the dividing issue of Lincoln's day, was a scourge, and is quite apparent from our vantage point and easy for us to pass judgment upon. Today, most people judge Lincoln as a great man, even some of my friends who live in South Carolina.

In 1860, on the other hand, half the people judged him as the devil incarnate, half the white ones anyway.

Judgment is not an exercise in determining an ultimate reality. There is no ultimate reality, only a reality from a certain point of view. The words *reality* and *relative*, are closely related because reality is a function of your relative place. You can say the sun rising in the morning and setting in the evening is a reality, an ultimate truth, but it's not. It's only a truth from the perspective of the surface of the earth. When Neil Armstrong flew from the earth to the moon in 1969, the sun did not set or rise for nearly a week as he left the earth's orbit bound for the moon. Once he was on the moon, however, it was the earth that appeared to rise and set on the horizon, just as the sun and moon did on the earth's surface. The sun rising is a reality based upon your relative position.

The creative process requires a shifting of viewpoint, different ways of looking at the same thing so as to determine strength and weakness. In the book *Six Thinking Hats,* Edward de Bono uses a hat metaphor as a clever and memorable way to describe this practice. For example, he calls a negative viewpoint a *black hat* and a positive viewpoint a *yellow hat*. Different personalities with different outlooks. These two viewpoints—positive and negative—are the ones you'll adopt to determine strengths and weaknesses. They're like the force that a structural engineer puts on a spacecraft in order to test it.

In the creative process, judgment isn't used to accept or reject ideas (that's decision making), it's a means to enhance ideas, a mechanism to drive the evolution of them, a way to manage the process.

The Role of Judgment

Brainstorming, a concept developed by an advertising executive, has given judgment a bad name in the creative world. In a brainstorming session, ideas are "not to be judged." Evidently, Alex Osborne, the advertising guy who created the concept, felt the critique of an idea would hurt feelings and so hinder the production or volume of ideas that surfaced in one of these sessions. People would be reluctant to contribute an idea for fear of its being rejected or ridiculed. So, he banned judgment from his creative process and by doing so he created a process that produced frivolous and trivial ideas. In advertising, it's not unusual for the frivolous to sell, so this probably worked for him. But in nearly every other domain, you're looking for ideas that have more substance. Ideas that solve important problems. So you need judgment—you need it to determine the quality of the idea and so drive the evolution of it. If you're worried about getting your feelings hurt, if your self-esteem is in thrall to the ideas you create, then you're in trouble. I say, "Grow up." As Harry Truman remarked, if you can't stand the heat, get out of the kitchen. Likewise, if you can't take the criticism, then put this book down and surrender any dreams of becoming more creative. Seriously.

Criticism is such an important part of the creative process that I've chosen to devote an entire chapter to it. Without exception, all of the creative people who have been profiled in this book were critics. They criticized the ideas of others, criticized their own ideas, and criticized themselves. Sure, the ego tends to be fragile, it'll get the best of you from time to time. But creatively successful people get over it, move on, and understand that criticism is needed in order to evolve theories, ideas, products, or anything else. It's the driving mechanism in the evolution of an idea.

The judgment of an idea does three things. First, it determines the weaknesses of the idea. Once they are identified, you can eliminate the flaws. Second, it determines the strengths of an idea. Once they are identified, you can enhance the strengths and make sure you don't eliminate them as you eliminate the flaws. Finally, and perhaps most importantly, you'll begin to develop a sense of the perfect solution, for it'll be an idea that has "these strong points" without "those weak points." In other words, judgment will develop your creative intuition. This allows you to know a good idea when you see it, the missing piece to your puzzle. Now you can describe the perfect solution and with this description go back to the second step, the borrowing step, and look for an idea that has the attributes you've described based on your judgments. As Jonas Salk said, "Intuition will tell the thinking mind where to look next." And intuition, my friends, is the result of judgment.

Judgment is a comparison. Whether you adopt a positive or negative point of view doesn't matter, you still need criteria to judge by. Put simply, the criteria are defined by the problem and its competing solutions. In business this means other products. In science it means other theories. In entertainment it means other performers. For Edison, illumination and safety were the criteria for the electric light. His comparisons were gas lamps, candles, and firelight. For Steve Jobs, applications and usability were the criteria for the personal computer. His comparisons were not other computers, but calculators, watches, and stereos. For Oprah Winfrey, entertainment was the criterion for her talk show. Her comparisons were soap operas, television movies, novels, as well as Phil Donahue. So, before you pass judgment, choose your comparisons wisely.

Negative Judgment

If you think a creative genius is always positive, optimistic, and encouraging, then you probably haven't met one. Creative thinkers are skeptics. The history of innovation could easily be called a history of skepticism. In fact, historian Jennifer Michael Hecht wrote such a book and called it *Doubt: The Great Doubters and Their Legacy of Innovation from Socrates and Jesus to Thomas Jefferson and Emily Dickin-*

son. Hecht argues that without doubt, without suspicion, it's impossible to progress. Positive thinking is closely related to contented thinking, happiness with the way things are. My golden retriever, Toby, is a very positive thinker, very happy with the way things are, but not very creative. Effective thinkers, on the other hand, are harsh critics, doubters, and proficient at negative judgment. Creativity is rooted in disbelief. Don't be misled by "creativity experts" who tell you negative thinking destroys creativity. It's actually the opposite, creativity can't exist without the negative.

In a 1997 interview with *BusinessWeek,* Steve Jobs said about Apple, "The products suck! There's no sex in them anymore!" Steve's never been one to mince words. And he's far from always being positive, optimistic, or encouraging. He's able to build great new products by seeing the problems with the current ones, even his own. Working for him, I'm told by friends who have, is a nightmare. He rips on people and their ideas with little or no concern for feelings or consideration. In meetings, he's the exact opposite of what Alex Osborne proposed with his so-called "brainstorming" concept. His typical response to a new idea is "That's shit." The joke at Apple was never to discuss ideas with Steve on an elevator because your chances of having a job by the time you reached your floor weren't very good. Employees started taking the stairs when he returned to the boardroom at Apple after his ten-year exile. People with fragile egos didn't work there. Not for long, anyway.

Yet, in 2008, *Time* magazine's editors named Steve Jobs one of the most influential thinkers on the planet. According to them, Jobs has changed the world. They said, "Jobs gets called mercurial, egomaniacal, a micromanager. If that sounds a little like a CEO doing his job, maybe that's because he is—and a mighty fine one." A popular biography on him is subtitled *The Greatest Second Act in the History of Business.* He has an innate sense of what will appeal to consumers. This is the guy who created the Macintosh, the iPod, iTunes, iPhone, and Pixar's *Toy Story*. He knows a good idea when he sees it. His creative intuition is finely tuned, a result, no doubt, of his legendary criticism of people, places, and things.

Actually, you—in common with most of us—are like Steve, for negative thinking comes naturally, even though you've been culturally trained to avoid it and to be positive. This is due, in part, to how your mind is wired. You're a pessimist by nature, for you're a descendent of negative thinkers; it was an important feature in the fight for survival. The paranoid survived, the one who assumed the rustling grass was a saber-toothed tiger, not the one who looked at the beauty of it as it swayed in the wind. Negativity is in your genes, so you might as well use it effectively. It's so natural that judgment itself is often solely associated with the negative. If my girlfriend, Deborah, calls me "judgmental," she's implying I'm being negative.

Negative judgment is a potent weapon. Remember, Nietzsche said that great intellects are skeptical. Don't be afraid to wield the negativity weapon with great force and precision. Without it, your ideas are frivolous and meaningless like those produced in an advertising brainstorming session. You destroy in order to create, or as Picasso said, "Every act of creation is first of all an act of destruction." It's the law of nature. The Tahoe forest is teeming with all sorts of life in various stages of creation and destruction. Old trees die and make room for the new ones. While the Giant Forest in Sequoia National Park (a few hundred miles south of Tahoe) is beautiful, there's not the same abundance of new life. The ancient trees are so old and so big they block the sun and make it difficult for younger trees to grow and so create a much starker forest, one devoid of new life. Creativity requires death: the death of an old idea. And it takes negativity to kill it, just as it takes the dead trees in Tahoe to make room for and provide the nutrients for the new ones.

When you analyze your creation, ask, *What are the negative attributes of this idea?* This isn't your opportunity to dismiss the idea; it's about identifying the explicit features that work to keep it from solving the problem. As you start to extend the metaphor you created in the third step, you'll discover places where it doesn't work. You'll use your negative judgment to identify those places and so abandon the metaphor where it's ill-conceived. Since the creative process is about constructing a solution out of borrowed components, eliminate the components that don't work. Isolate them and put them into the "weakness" column so you can come back later and get rid of them. Be specific and dispassionate. This is a technical matter, not an emotional one. Your judgments are based upon the criteria you've established and the comparisons you've chosen. In other words, your judgments are based upon the problem you identified in the very first step of *Borrowing Brilliance*.

Imagine Masaru Ibuka as he examined the first prototype of his Sony Walkman in 1978. The problem he was solving was a *personal entertainment system* one and he was taken aback when his engineers showed him the first large, bulky working model. He shook his head in disgust. Examining it more closely, he realized that the model still contained a "recording" element. This wasn't surprising, since the engineers worked for the Tape Recording Division of the company. But Ibuka was conceiving this product differently—his primary criteria were "small" and "high fidelity." Anything that didn't contribute to these attributes was evident when Ibuka put on his "negative" hat. He was comparing his new product to a transistor radio or a dime-store paperback novel. So, he chastised the engineers and demanded they remove the recording elements.

Steve Jobs treated product designers and engineers at Apple to even harsher judgment. "That's shit" was repeated over and over by Steve. He had a clear vision of the Macintosh and the problems it was trying to solve. He summarily rejected any element, big or small, that didn't contribute to solving these problems. He blasted the early Mac prototypes. According to writers Jeff Young and Bill Simon, in their biography of Jobs called *iCon*, "One day Steve went into a design meeting with a telephone book, and threw it on the table. 'That's how big the Macintosh can be. Nothing bigger will make it. Consumers won't stand for it if it's any larger.'"

"Hey, and there's something else," he said flippantly before he walked out, "I'm tired of all these square, squat boxy-looking computers. Why can't we build one that's taller, rather than wider? Think about it."

Unlike his competitors, Jobs used more than computers for comparison. He was using stereos, fine watches, drawing pads, desktops, and Japanese artwork to judge his product. Young and Simon said, "What made the job so frustrating for the team was Steve's apparent total inability to overlook any detail of the project. He was a micromanager to the nth degree. He cared passionately about the smallest of items. Eventually the final result was better for it; however, the path was tortuous." He controlled his vision and the evolution of the product through harsh judgments.

Of course, I'm not advocating the mistreatment of your business associates in the quest for a creative idea. I only use Steve Jobs as a blatant example of effective negative thinking. He's so obvious about it and he's a creative genius, a creative role model, a man with impeccable judgment. However, this doesn't mean you need to be cruel or unkind. With others, I've learned to soften the blow by borrowing a concept from the Roman Catholic Church.

The Devil's Advocate was established in 1537 as a title given to a canonization lawyer who argued against the God's advocate in the formal electoral process designed to determine sainthood. The Devil's advocate didn't necessarily believe in the unworthiness of the candidate but was asked to take the view of Satan to see if God's advocate could properly defend the candidate against these attacks. I'm sure it wasn't considered a plum assignment at the Vatican, having to dish against John the Baptist, mock the Apostle Paul, or describe the shameful nature of Mother Teresa. You can imagine the conversation in the Vatican locker room after one of these debates, "Hey Father Murray, nice job making Mother Teresa look like a common harlot out there today. Hell of a performance!" Seriously, though, it's the perfect metaphor for softening the critical blows to others in the creative process. By telling someone, before a negative judgment, you're playing "the devil's advocate," you're stating that this is a viewpoint, not necessarily your belief or ultimate judgment of the idea. Everyone understands the metaphor and it can save fragile egos. (Of course, anyone who can't take constructive criticism isn't going to get very far in the creative process. But, whatever.)

In the years I spent constructing Preferred Capital, I personally trained more than five hundred salespeople. It was a critical part of my business model, and so I developed a highly structured and highly successful sales training program. While many factors determine the success of a sales rep, the ability to identify and overcome objections is the most important, and the most misunderstood. To someone unfamiliar with selling, it seems counterintuitive to seek out objections when trying to close a deal. A new salesperson wants to avoid or not look for them. But overcoming objections is a key part of the sales cycle, and salespeople have developed clever techniques to do this. For example, a *trial close* determines the objections, as when a car salesman asks, "What color do you want your Prius in?" He hasn't asked for the order, he's "trying" the close, putting the sale under some load.

So, using negative judgment is thinking like a salesman and asking, *What are the potential objections to this idea?* Imagine your idea out in the world without you. What are the objections people will have to it? It cannot overcome these if it doesn't know what they are, just as a climber cannot scale a mountain unless he sees the cliffs, or a salesman cannot close a deal unless he sees the objections. Every idea will have barriers; you need to figure out what they are so you can get around them. That's why you need negative judgment in the creative process. It's why, in Steve Jobs's defense, he says . . . "That's shit."

Positive Judgment

While the negative viewpoint is necessary to identify barriers, it must be tempered and combined with the positive view or else the negative becomes completely destructive. Positive judgment determines the strength of your idea. In the forest, the old trees die to make room for the new ones. Positive judgment is the fuel that provides the intellectual material for growth. Optimism identifies the opportunities, the clear passageway to the summit.

So, while at times Steve Jobs can be mean-spirited, pessimistic, and destructive, at other times he can be incredibly optimistic, complimentary, and productive. For instance, he described a new operating system by saying, "We made the buttons on the screen look so good you'll want to lick them." Young and Simon call this a "stunning contradiction in Steve Jobs's personality." It's not really stunning, though. It's just the other viewpoint that every creative genius knows is necessary to grasp the opportunities that lie ahead in the evolution of an idea. It's putting a different intellectual load upon the creation. Jobs does this, it seems, subconsciously. It's part of his nature. For you and me it's got to be a conscious effort, deliberately changing the vantage point from which you judge your creations. You'll say to yourself, *Okay*, now *let's put the positive hat on and see what's of value in this idea. What piece of this idea can I use in the next iteration of this solution? What should survive?*

"Steve has a power of vision that is almost frightening," said Trip Hawkins, an early Apple marketing manager. "When Steve believes in something, the power of that vision can literally sweep aside any objections, problems, or whatever. They just cease to exist." In other words, Steve can grab hold of something, sometimes just a tiny piece of an idea, and rally behind it with the knowledge that he'll figure out how to solve the problems that surround it later. He can see the positive, a pathway to the summit, in a mountain of negative. Others see the troubles. Steve sees, or senses, the solutions. Apple succeeded, Hawkins says, because "we really believed in what we were doing. The key thing was that we weren't in it for the money. We were out to change the world." Or as Steve so positively stated, "We'll make it so important that it will make a dent in the universe."

As a creative thinker, you'll want to develop a similar contradictory personality. It's the contradiction that makes you confront the various viewpoints that will put your solution under intellectual pressure, as a testing engineer does, in order to determine strengths and weaknesses. Always being negative is just as creatively dangerous as always being positive. You need to be both. An idea evolves by eliminating negative things and enhancing positive things. You need the "stunning contradiction in personality."

It's difficult to be positive, though. It's not as natural as the negative. Positive judgment requires imagination that will let you see something valuable among things of little value, the way a sculptor sees the form of a statue inside a block of rough marble. Michelangelo needed this eye to see the lifeless body of Christ in the stone before he created *The Pietà*, his masterpiece. Positive judgment is a constructive point of view.

It isn't always easy. Even great ideas are sometimes hard to see. When Darwin published his theory of natural selection it didn't make the front page of the London *Times*. In fact, it didn't even make the front page of the Royal Society newsletter in which it was first published. It was somewhere in the middle, buried among ideas that have long since gone extinct, even though Darwin was a respected naturalist and a famous author. Of course, this quickly changed and Darwin's theory became well known in his own day.

At the same time, in Austria, a young monk was making equally important contributions to evolutionary biology by single-handedly creating the field of genetics. But he struggled in obscurity, going completely unnoticed and dying in the dim shadows of his time. Decades later, Gregor Mendel's work was recognized for its revolutionary and creative value. No one, perhaps not even Mendel himself, understood the significance of his ideas. Seeing the value of an idea isn't always easy. It takes a finely tuned positive viewpoint.

So, as you analyze your creation, ask, *What are the positive attributes of this idea?* This isn't about whether you like the idea or not, it's not about the acceptance of the idea, it's about identifying the explicit features that work to solve the defined problem. Remember, the creative process is a matter of combining different components to create a new structure. By adopting this positive viewpoint, you're identifying those components you find helpful in achieving your goals. Isolate them and place them into the "strength" column so you can come back later and use them to construct the next generation of solutions. You don't want to eliminate your strengths on the upcoming step. Make sure you've acknowledged what they are. Be specific.

Think of the positive as a means to look for opportunity. Just as the smart climber looks for the barriers, those places too dangerous to climb, she also looks for opportunities, those places with an opening or safe opportunity to pass. She scans the mountain above and looks for a gap, a notch in the peak, a weakness in the mountain. The same is true for you and the judgment of your ideas. You need to sense the opportunities that lie in the idea, that lie in your subject, even if the idea itself is not worthy of implementation. Most ideas have possibilities, the seeds of another idea, within them. Put a load on the idea and see if it has any strength. The structure may fail, but there may be a part of it that's strong, that can be used elsewhere. That's an opportunity. For example, you might design a new product that's too expensive to construct, and so too expensive to buy. This should send you back to the drawing board. But before you begin again, see if there's a component to this design that's inexpensive. If there is, use it to begin your next design. In other words, before you throw away your idea, see if there's anything in it worth keeping.

Positive thinking is constructive, for it illuminates possibilities and opens doors. However, by itself it's ineffectual, for it inhibits the awareness of obstacles. Negative thinking, on the other hand, though useful in revealing objections, illuminating barriers, and identifying pitfalls, by itself tends to destroy creative thinking, for it's discouraging. Taken together, however, positive and negative thinking become valuable tools that help your ideas to evolve by revealing a pathway—the open doors and the walls that define those doors. And they come together most effectively in "the creative debate."

Socrates believed that learning occurred when a teacher presented a concept, the student challenged it, and the teacher replied to the challenge. Together, teacher and student created the learning. Ultimately, the teacher developed a deeper understanding of the concept or abandoned it if he couldn't defend it. Through this process, ideas were perfected or discarded for new ones. Today, this is called the Socratic method. It isn't used in the modern classroom but it's a form of debate you can use in the creative process. In fact, in business, or any domain that requires creative collaboration, debate is the way to combine positive and negative viewpoints. You attack and defend your ideas. You should discard those ideas, or the pieces of those ideas, you can't defend. You should move ahead with those ideas, or pieces of those ideas, you can.

"Steve Jobs liked people who had the guts to stand up to him," his biographers Young and Simon said, "but with a very demanding limitation: This applied only to people he respected, people who had a real contribution to make and whom he could look on in some respects as his equals." Put simply, this meant people who could defend their contrary positions through logical argument and with evidence to support their claims. Disagreeing without a clearly defined position, at Apple, was tantamount to committing career suicide.

As a creative thinker, you develop these skills through logic and evidence, not to win arguments, but to advance your ideas, evolve them, perfect them, or abandon those you can't defend. Debating adds depth to your thinking and is the perfect venue and meeting place for the positive and negative viewpoints. A formal debate consists of four things: a proposition (your idea); issues (unsupported claims); arguments (claims supported by logic); and evidence (facts used to support claims). Understanding these things develops the structure of your thoughts. Study them. Debate yourself. Take both sides of the issues. Argue. Challenge. Defend. Ask yourself: *What are the issues, arguments, and evidence that work to endorse my idea?* And then ask: *What are the issues, arguments, and evidence that work to damage the viability of my idea?* Answering these questions illuminates the strengths and weaknesses of your idea. Creative debate is a way to think; it is not necessarily the formal act that you might imagine goes on in the stuffy halls of Harvard University. It adds a depth to your thinking that enhances your ideas.

Like Apple, Microsoft also encouraged debate. In the Gates biography *Hard Drive*, the authors say, ". . . they expected to be able to challenge Gates. In fact, he wanted them to argue with him. His confrontational style of management helped Microsoft maintain its edge, its mental toughness. It made those who worked for him think things through. These are qualities that continue to distinguish Microsoft to this day. It is a culture that never gives employees a chance to get complacent because as they do, someone is going to challenge them. Gates was not afraid to change his mind if someone made a convincing argument."

Steve Wood, one of the first programmers to be hired by Microsoft, added, "Bill is not dogmatic about things. He's very pragmatic, he can be extremely vocal and persuasive in arguing one side of an issue, and a day or two later he will say he was wrong and let's get on with it. There are not that many people who have the drive and the intensity and entrepreneurial qualities to be that successful who also have the ability to put their ego aside. That's a rare trait." And it's a trait that you and I need to emulate. It's the only way to drive the evolution of an idea.

A word of caution. Winning an argument with someone doesn't mean your ideas are sound. Victory is prejudiced by factors other than logic and reason. Steve Jobs can easily out-argue me or refute my ideas in a board meeting, panel discussion, or on Larry King by virtue of his reputation, position, and personality. Likewise, my daughter can beat me in an argument by simply turning on the tears. He who shouts the loudest or cries the hardest can often triumph in a debate. So beware, this doesn't mean the idea, or pieces of it, aren't valuable just because you've lost an argument.

To sum up, by utilizing positive and negative judgment and the debating process you determine your idea's weak and strong points respectively. These opposing forces enhance your idea and help develop your creative intuition, your sense for the perfect solution, one that'll have "these strengths" without "those weaknesses." Before we explore intuition in more depth and then conclude this chapter, let's take a quick side trip to make sure you're headed in the right direction.

Emotional Judgment

Everything you do in the creative process is built upon a foundation of problems. Once defined in the first step, your problem draws a map. In the second step, this map is used to find the borrowed materials for the construction of your idea. In the third step, the problem glues together the two ideas and forms the metaphorical structure for your solution. And in this step, the problem provides the criteria and comparisons that you judge your ideas upon. But the problem is elusive. You may have defined it incorrectly in the first step, and if you have, everything else you do is ill-conceived. So, I've developed a checkpoint in the creative process, a quick test to make sure you're on the right path. I call it "emotional judgment."

Positive and negative judgment are consciously taken viewpoints. In other words, before you pass positive judgment you consciously put yourself in an optimistic frame of mind by asking yourself to describe how this idea best solves the identified problem. With the negative, you ask yourself how it doesn't work to solve the problem. Then you debate yourself, or others, taking both the positive and negative positions. These conversations will be steeped in logic, reason, and hard evidence. This debate will lead to some kind of overall opinion about the idea. Positive or negative will dominate. The idea, on the whole, will be either good or bad. It'll "need a lot of work" or "almost be there." This estimation is the result of logic and reason, so I call it my "logical opinion."

Now, before you move on to the next step, before you begin enhancing or evolving the idea, you need to take off your logical hat and put on your emotional one. The logical hat, the positive/negative one, requires well-formed arguments and evidence in the form of hard facts and figures. The emotional hat, in contrast, requires no argument, no evidence, and no hard facts to support it. The emotional hat is based on a very simple question: *How do you "feel" about the idea?*

The emotional viewpoint is the viewpoint of the subconscious mind. In other words, when you ask yourself, *How do I feel?*, you're posing that question to your shadow self. You're asking your subconscious mind to pass judgment on the idea and answer you in the form of a feeling. Remember, emotions are the language of the subconscious. So, using the techniques from the previous chapter, you'll put a short pause in your thinking before and after you ask the question. You'll want to clear your mind of the deep grooves that the positive/ negative debates have dug into it, for they'll influence the subconscious if you're not careful. That's why I like to insert a time gap to separate the emotional judgment from the logical judgment. The answer will be either a positive feeling, a negative feeling, or an indifferent one. That's my "emotional opinion."

Once I've got it, I take that feeling, my emotional opinion, and compare it to the logical opinion I've constructed through the debate of positive and negative judgments. If they're consistent with each other, I conclude I'm on the right path. However, if they're inconsistent with each other, one positive and one negative, then I've got trouble and I conclude that I may be on the wrong path. I conclude that I may be solving the wrong problem, that I may be building my idea on a foundation of sand.

Let me explain this process with a practical example. While I was working on the TurboTax direct marketing program, Tom asked me to review the new advertising program his marketing department and advertising agency were constructing. Tom was about to launch the first television commercials in the history of TurboTax, an investment of more than thirty million dollars. He wanted my opinion about the commercials created. He wanted everyone's opinion. So, we crowded into a conference room and the advertising agency presented half a dozen different ads, using storyboards and acting out the different options. Each commercial was very clever. Funny. Just what you'd expect from a world-class advertising agency and a high-equity brand like TurboTax. Each commercial was consistent with the marketing brief that we'd provided the agency. A brief describes the target customer, our position in the market, and the features and benefits of using our product. The primary feature of our product was its accuracy. It was tax software, after all. When the advertising agency finished, they asked us to rate the commercials so we could choose the one to put our thirty million dollars behind.

As we went around the room, each executive had a well-thought-out, logical reason why he or she liked one commercial over the others. Using the marketing brief as the criterion on which to judge the commercial, the executives, through debate among themselves and the agency people, zeroed in on a specific commercial. Everyone seemed to like the second one. Finally, it was my turn to give my two cents.

Tom asked me, "Dave, you've been quiet so far. What do you think? Which commercial do you like best and why?"

"I don't like any of them," I said. The room went silent. Everyone turned toward me.

"Okay," Tom said, "what's the problem, then?"

"I don't know. These just don't feel right to me," I answered. You see, by the marketing-brief criteria, the commercials were very good, clearly and cleverly stating our position and primary benefit. In fact, using logic and reason I had come to the same conclusion as the rest of the group and felt the second commercial was the best. But my emotional opinion was different from my logical opinion. My emotions didn't like any of them.

"Can you be more specific?" Tom asked.

"I don't know, Tom. Something's wrong but I can't put my finger on it," I answered to the chagrin of the advertising agency and some of the other executives. It didn't matter, though. I was only a consultant for the direct marketing program and there were a lot more important people in the room, including an Intuit board member. The meeting continued on and the agency was told to work on the second commercial, but with some adjustments and changes from the group's debate. I went back to my office.

I thought about it all afternoon. As I sat and pondered, it finally hit me. It wasn't the commercial I disliked for it was very good. It was the marketing brief and the designation of the primary feature— "accuracy"—that was the source of my negative feelings. It was in every one of the commercials, for that was the direction we'd given to the agency. So I went to Tom to ask why he had chosen "accuracy" as the commercial's primary attribute. He told me to go see the research department. They showed me reams of data. Surveys, focus groups, and interviews with customers who all agreed that "accuracy" was the number one trait that users expected in a tax software program. It was ranked higher than simplicity, usability, and speed of use. Much higher, as I recall.

"Aha," I said as I looked more deeply at the data. I went back to Tom.

"You're solving the wrong problem," I told him.

"What do you mean?" he asked.

I said, "First of all, none of your surveys, focus groups, or interviews, as far as I can tell, are asking customers what feature they use to make their buying decision. They've all been asked to rank the important features of tax software, not what features influence them to purchase. It would be like asking diners to rank important features of a nice restaurant. I am sure that 'cleanliness' of the kitchen would be very, very important. Maybe number one. But I wouldn't design my advertising program around it. Few people use the health inspector's report to make a dining decision. They expect a clean kitchen, and a restaurant that advertised a clean kitchen would seem very suspicious."

I went on to tell Tom that accuracy in tax software is like cleanliness in a fine restaurant. I had a hunch that people expected it and if we started advertising it we might destroy that expectation. A follow-up investigation by the research department confirmed the hunch. So, TurboTax dropped the "accuracy" feature from all of their commercials and focused on speed and simplicity.

The reason I hadn't liked the commercial was because it was solving the wrong problem. My logical opinion told me the ads were very good and consistent with our marketing brief. However, my emotional opinion told me something completely different. When this happens, I've discovered, it usually means that I've defined the problem incorrectly. The TurboTax commercials were being built on a foundation of sand.

I've also discovered that voicing my emotional opinion can create discomfort in groups, especially in business situations. Business is, supposedly, based on logic, not feelings. So it's difficult to express them without including the logic behind them. When I told the Intuit executives that I didn't like the idea, but didn't know why, they dismissed the opinion because it wasn't backed by logic and reason. A few people rolled their eyes.

So, I use a device I call my "spider-sense" to introduce my emotional opinion. It's like using the devil's advocate to introduce negative judgments. Now, the exchange goes like this.

"I don't like the idea."

"Why?"

"My spider-sense tells me we're thinking about this the wrong way."

Much to my astonishment, this works. My feelings are more apt to be accepted without logic and reason with this simple introduction. Everyone realizes I haven't fully formed my opinion, that there's an issue that hasn't been uncovered yet and is going to take more time to surface. It brings emotions into a situation that's typically uncomfortable with them.

To use it, refer to your emotions as your "spider-sense." Comic book superhero Spider-Man is the alter ego of Peter Parker, who was bitten by a radioactive spider and gained spiderlike attributes, such as the ability to climb walls and spin webs. He also has a heightened sense of perception. He can detect the slightest change in a situational pattern, like the delicate change in air pressure that a hidden criminal creates in a room; his *spider-sense* says something's wrong. He can sense and anticipate danger without logic and reason. Of course, this is a great metaphor for emotional judgment, for it can create a heightened sense of awareness too. The ability to sense danger in an idea or to sense an inconsistency in thinking is a powerful tool for the creative thinker. You've spent so much time building your solution on a foundation of problems, you need to periodically check to make sure the foundation is solid.

Intuition as a Result of Judgment

Now that you've learned about positive, negative, and emotional judgments you're ready to move on to the next step. You're ready to begin evolving your idea by eliminating its weakness and enhancing its strength. And as you enter this stage you bring with yourself a more finely tuned sense of creative intuition. Your positive and negative evaluations have crafted this intelligence. Intuition is the magnificent side effect of judgment.

In 1979, as legend has it and according to Young and Simon, Steve Jobs and six other Apple employees managed to get a tour of the highly secretive Xerox Palo Alto Research Center (PARC). It was at the forefront of advanced computer technology research while Apple was just a start-up owned by a couple of pot-smoking hippies with a single, simple product. According to Larry Tessler, the PARC scientist giving the tour, he thought that "these were a bunch of hackers, and they didn't really understand computer science. They wouldn't really understand what we were doing, just see pretty things dancing on the screen."

However, Tessler witnessed just the opposite. The Apple crew was intense. They asked good questions and easily grasped the concepts behind the answers. Tessler said, "And Jobs was pacing around the room, jumping up and down and acting up the whole time. He was very excited."

"Then when he began seeing things that I could do on-screen, he watched for about a minute and then he was jumping around the room, shouting in the air, saying, 'Why aren't you doing anything with this?! This is the greatest thing! This is revolutionary.'" Tessler explained. What Steve was looking at was a screen with a mechanical pointer that the user could manipulate to select various on-screen objects that represented different tasks. You see, Steve was looking at the prototype of a mouse and what you call desktop icons or what a software designer calls a graphical user interface (GUI). The operator could open different programs at the same time using different "windows." The mouse was nothing new. It had been developed more than a decade earlier. On-screen graphics weren't new. And neither were multiple program operations. What was new was how all of these things worked together, how the combinations complemented each other and created a whole new computer user experience.

Tessler said, "Nobody else who had ever seen the demo cared as much about the subtleties. Why the patterns were there in the title of the window. Why the pop-up menus looked the way they did." He continued, "What impressed me was that their questions were better than any I had heard in the seven years I had been at Xerox. From anybody—Xerox employee, visitor, university professor, student. Their questions showed that they understood all the implications, and they understood the subtleties too. By the end of the demo I was convinced that I was going to leave Xerox and go to Apple." Which he did.

This story epitomizes creative intuition. Larry Tessler had given the same demo to hundreds of other people. Presumably, very smart people. But it took a twenty-five-year-old entrepreneur to see the revolutionary potential in this combination of technologies. The question for us is why did Steve Jobs have this creative intuition when the hundreds of others who saw it did not?

The answer lies, I believe, in what Young and Simon call the "stunning contradiction in Steve Jobs's personality." His intense judgment, both positive and negative, creates a vision of the perfect solution: It's one that has "these strengths" (from his positive personality) and doesn't have "those weaknesses" (from his negative personality). When he saw the Xerox demo he recognized it as the perfect solution for the personal computer revolution he was leading. It had the necessary strengths (simple to use, intuitive, and fun to interact with) and was lacking an important weakness (having to memorize cryptic commands to operate a computer). You see, Steve was solving a different problem than were most of the others who had seen Tessler's demo. He wanted to make computers easy to use, accessible to the masses, make them "personal." His intense judgment of computers, using this criterion, caused him to recognize the problem with command-led operations. Others didn't see this as a problem because they were mostly programmers and able to memorize the commands. Casual users could not. Steve didn't know how to solve this problem but he was well aware of it, either consciously or subconsciously. So when he saw the mouse and GUI he went, "Aha, that's it. That's the solution to the problem!"

For Steve Jobs, this is an innate ability. It's part of his makeup. He subconsciously shifts between positive, negative, and emotional judgments and creates a keen awareness of problems and their solutions. He knows a good thing when he sees it. He did the same thing with the iPod. He didn't invent MP3 players, they already existed. What he did was to combine them with an online shopping experience, iTunes, and so create a seamless experience so simple and powerful that it changed how we interact with music. He did the same thing with Pixar. He didn't create the technology, George Lucas and his team did. However, he sensed the revolutionary nature of the Pixar software and combined it with world class storytelling in the person of John Lasseter (Pixar's chief creative officer). Investing less than ten million dollars to buy the company, in a few years he turned it around and sold it, through an IPO, for billions of dollars. Again, these contradictory elements in his personality allow him to hone his creative intuition to continuously innovate and make more "dents" in the universe.

You and I will probably never put a "dent" in the universe. We'll never have the same sense of creative intuition that Steve Jobs has, but we can develop our intuition nonetheless. You can simulate his way of thinking even if you can't acquire his innate abilities. You can think like him. You can adopt these contradictory personalities by consciously assuming these various points of view, what de Bono calls "thinking hats."

In the final step of *Borrowing Brilliance*, you'll generate a lot of different ideas. Then you'll apply different loads to them, positive and negative viewpoints, that will help you construct a detailed list of positive and negative attributes for the perfect solution to your problem. The more ideas you have, the more judgments you pass, the finer-tuned your intuition becomes. Eventually, you'll become more like Steve Jobs and know a good idea when you see it. Or, more importantly, you'll know where to go to seek out the components you need for your solution. As Jonas Salk said, intuition will tell a creative thinker where to look next. So, after you construct your list of positives and negatives, ask: *Where can I go to get the kind of stuff that I'm describing?*

Creative intuition isn't as magical as it seems. It's simply a by-product of analyzing your ideas with positive and negative judgments.

• • • •

I've never liked the brainstorming process and now I know why—the elimination of judgment produces ideas that are silly, lighthearted, and useless. I've come to realize that without skepticism and negative thinking the creative process is doomed to failure. This is contrary to what I've learned. I need to test my ideas the way I tested spacecraft. I also realize that it's judgment that creates intuition and aids me in the search for idea components to borrow.

I'm beginning to understand this stuff.

The Fifth Step in the Long, Strange Trip

Tom says he's got a position for me at Intuit. I'm skeptical. Even though I've had my ass kicked in the small-business entrepreneurial world, it still feels like the right place for me. I like working for myself, with my own ideas, and seeing if I can make them happen. I can't imagine commuting to the office every morning in hopes of making my bonus at the end of the year. Answering to a boss. Or Wall Street.

"You'll be the head of innovation," he tells me.

"Really. Hmm." *Interesting*, I think. I ask, "What's that mean?"

"You'll be the idea guy at the company. You'll come up with new ideas and help other people to come up with them too," he says.

I'd never heard of such a thing. Tom tells me that Scott Cook, the founder of Intuit, plays that role at the company and wants to see if he can expand it. He wants to build an Innovation Department, a group of people dedicated to thinking and driving innovation into the organization. He wants to know if I'm interested. I say yes. That sounds pretty cool.

So I sit down to talk with Scott. He's familiar with the TurboTax MyCD program. He tells me he loves it, such a simple idea, he says, so obvious, but no one ever thought of it.

"How'd you think of it?" he asks.

I tell him the story. How I defined the problem. How I gathered the materials. How I made the combinations. And how it popped into my head while I was climbing San Jacinto. I also tell him how I know Tom. How I started my own company. How I grew it. How Tom offered me twenty-five million dollars for it. How the bank offered me fifty million. How I took the bank's offer. How the bank went under. How I went under. And how I ended up hidden in a small apartment in Tempe, Arizona. I don't tell him I had become an alcoholic; he could figure that out on his own.

"Wow," he says. "That's quite a story. You should write a book."

"Yeah, right," I answer.

So I become the head of innovation at Intuit. It's my job to come up with new ideas and to help others to come up with them too. I create a series of presentations I give to Intuit employees—an Innovation Training Program. I'm being paid to think about thinking and I like it. I read more about creativity as a subject. I talk with Scott. With Tom. Examine my ideas. Examine other people's ideas. I study Newton and Darwin more closely, looking for clues into their process. I start developing thinking tools, using a concept I borrow from Don Norman in his book *Things That Make Us Smart*. They're mental artifacts, things used to enhance creative ability. I teach these tools to Intuit employees and in return they teach me more. I'm constantly sorting and grouping these tools as I acquire them.

A year goes by. Tom and I become friends. He asks if I'd ever be interested in starting a company again. I say I would. "So would I," he says. "We should start looking for some opportunities. Ideas for new companies," he says.

"Actually," I say, "there's a great opportunity I've been thinking about."

"Really. What's that?" he asks.

I tell him the opportunity I see. He smiles. He tells me he was thinking the same thing. We agree to talk more about it later.

CHAPTER SIX

THE SIXTH STEP—ENHANCING

TRIAL AND ERROR AS THE PASSAGE TO THE CREATIVE SOLUTION

Traveling back in time two years, I see myself at fourteen thousand feet, at the top of the Mountaineers Chute on Whitney, the tallest mountain in the continental United States. Below me is Owens Valley, the deepest valley on the continent, and over the next ridge is Death Valley, the lowest point on the continent. It's extreme here. I'm only a few hundred feet from the summit, but I have a problem. I'm lost.

John Muir, founder of the Sierra Club, was the first person to climb the eastern escarpment of Whitney, up the same route I'm on. I think he was insane to do it. It's scary up here. There's nothing but rock, rock and extreme exposures, peaks and valleys, couloirs and cliffs. I have just ascended a two-thousand-foot couloir, a rock chute, a scar in the dominant front face of Whitney. The guidebook tells me to walk through the notch at the top of the chute, turn left, and then scramble up the north slope to the summit. I do, but when I look up my heart drops. This can't be right. Too steep. Too scary. And too exposed. Below me is a thousand feet of nothingness, a drop to the valley floor. *This is bullshit,* I think to myself. I wait for my friends.

Meyers is the first one up. Next Ryan. Then Mike. Meyers looks up and laughs; he's got a crazy look in his eye. Ryan looks up and mutters, "Screw this bullshit," to himself; he's pissed off. Mike doesn't look up or say anything; he's too tired to care either way.

"What do ya think?" I ask.

"I think we oughta get our butts on the summit," Meyers says.

"I think we need ropes and helmets," Ryan says. He points up. About a hundred feet above us is another climbing party. They've got ropes and helmets. And they still look unsure.

"We don't need no stinking ropes!" Meyers says. "Just don't look down. And don't fall! Come on. Let's go, girls!"

Meyers is the same friend who talked me into climbing McKinley. The same friend who talked me into paddling across a crocodile-infested estuary at Witches Rock in Costa Rica. He hadn't killed me—yet—but it wasn't for lack of trying. Of course, he expected me to lead us up to the summit, for I'd been taking rock climbing classes all summer.

I begin walking along a wide ledge, looking for a place to start climbing, trying to pick out a route, a safe passage to the top. I don't see one, so I keep walking. After several hundred feet of ledge, it comes to an abrupt stop. I'm on a deadly cliff, so I turn and start mountaineering up. I'm not sure if this is the way John Muir went. I'm not sure I'm on route.

At first, the climbing is straightforward, easy, with plenty of handholds and footholds. I feel good. My rock-climbing class gives me confidence. Ryan and Mike, on the other hand, are struggling, unsure of themselves. I think this isn't a good place to be learning how to climb, especially without ropes. Meyers is laughing again.

Then I run into the first dead end. I have climbed into a slot that ends in a flat slab of steep rock, too steep to climb without gear. A fall here means death. I'm stuck. I climb down to a ledge below. I need to find another way. I must be off-route.

I start up again along a different way. I push upward and then—Holy shit!—the small rock I'm standing on lets loose. I start to fall but I grab on to the rocks above my head. I catch myself. I reach for another rock, test it, and it's loose too. *This is bullshit,* I think to myself again. It's bad enough being up here with all this exposure, but now I can't trust the rock I'm on, it's soft, it's letting loose, and I'm dropping bombs on my friends below. I hear Ryan cussing beneath me but can't see him. I dislodge a loose rock and listen to it as it skips along the cliff and then . . . nothing . . . I never hear it land. This sucks.

I keep climbing, though. No choice at this point. Once again, however, I find myself boxed into a slot, no place to go, like I'm a laboratory rat searching for cheese in a maze. Dead end. I have to down-climb, again. I try another way. This is like Pan's Labyrinth. I climb up. Get stuck. Climb down. Drop a few rocks on my friends. Climb up again. And I get stuck again. Dead ends that lead to other dead ends. I'm tired and scared. Meyers isn't laughing anymore. We're all tired and scared. We're lost. I know we're off-route.

Finally I make my way to within forty feet of the summit. I hear people above me. Laughing. Having lunch. But I'm trapped. There's a solid wall of rock blocking my way. I lean out and look across the void. About a hundred feet to the north is a break in the wall, a weakness, a door to the summit. The only problem is I have to tightrope-walk across a small ledge, no wider than my fist, to get to it. And I can't trust the rock. Oh, well. I guess I gotta do it. There's no doubt we're off-route. This isn't in the guidebook. I go for it.

A few minutes later I'm on the summit. I look down and yell to

Meyers, Ryan, and Mike. I tell them to go left. Then right. It's easy to see the route from here. It's easy to see all the mistakes I made trying to get us up.

A few minutes later I lean over and give Meyers my hand and pull him up to the summit plateau. He looks at me—laughs—and says, "John Muir is fugging nuts!"

· · · ·

Two years later, this story pops into my mind as I build the Innovation Training Program at Intuit. Steve Bennett, the CEO, is a General Electric alumnus and well-versed in Six Sigma and process management. He wants a process—a step-by-step procedure for developing new business innovations. So I ask myself: *What are the steps to constructing a creative idea?* Instead of getting an answer, this story comes to me. It's not hard to interpret. My shadow is saying the creative process is like climbing the Mountaineers Chute on Whitney, it's a matter of trial and error. Ideas are illusive, like the summit, and finding your way is dangerous and marked by failure. This isn't the answer Bennett wants, so I give him a six-step process. He likes it even though the sixth step is really an iterative one. It's the first five steps repeated and it fixes or enhances the ideas from the initial steps.

Of course, I should apologize for misleading you just as I misled Bennett. The creative process is not really six steps, it's actually a self-organizing process, a haphazard one, more circular than it is linear. It works this way because this is the way all organic systems construct things, this is the way your mind is constructed and does its own construction, for it's inherent in the biology of how you're wired.

Let me explain.

Left-Brain Thinking

In 1963 a young college professor at Caltech began a series of experiments with half a dozen epileptic patients who had undergone radical brain surgery to try to alleviate intractable grand mal seizures. This surgery severed the corpus callosum, the dense nerve connections between the right and left hemispheres of the brain. Roger Sperry gave these patients simple tasks to perform, like naming objects or grouping things together, while covering the right eye and then switching and covering the left. Since the right eye connects to the left half of the brain and the left eye to the right, Sperry found that certain tasks could only be performed by the left brain and others only by the right. He won a Nobel Prize in 1981 for this work.

While not absolute, Sperry found that the left hemisphere specialized in reductionist tasks like language, mathematics, and categorization and the right hemisphere performed holistic tasks like determining shapes, patterns, and spatial perceptions. In other words, the left brain takes things apart, perceiving the pieces, and the right brain puts them back together, perceiving the whole. If you're right-handed, as most people are, then you're dominated by the left brain. You're good at taking things apart but find it more difficult to put things back together and make out the whole. You live in a world dominated by left-brained thinking. A world of details.

Imagine someone asking you to design and construct an earth-orbiting space station. Where would you start? This was the question the Johnson Space Center asked my boss Bob Pedralia in 1985. The process we went through to answer this question was one dominated by left-brained thinking. The first thing Bob did was to break the concept into its component parts, to reduce it to its pieces. I recall them being: structural, navigational, propulsion, life support, energy, and habitation. Each piece was given to a separate team. The team, in turn, broke their piece into its component parts, the pieces into pieces. For example, the structural team broke itself into the truss structure, the module structure, and the docking structure. The truss structure was broken into geometry, materials, and fasteners. The fasteners reduced to welds, bolts, screws, and nuts. And so on for each piece.

Creative thinking requires left-brained thinking because the construction of an idea is the act of combining borrowed components to form the structure for a new solution. You need left-brained-reductionist thinking to perceive those components. Most people are pretty good at this. The entire educational system is based on left-brained thinking. A world of details.

However, taken to the extreme, you become lost in the details and become ignorant as to how your idea fits into the overall structure of the solution. You can't see the forest through the trees. Instead, you're lost studying an acorn or a leaf, and have no idea where you are in the forest, or even that the forest exists. While climbing the north slope of Whitney I had to keep looking up, assessing my situation, and then refocusing on the rock in front of me. Reductionist thinking, alone, is dangerous and ineffectual. It must be balanced with holistic thinking. With right-brained thinking.

Let me explain.

Right-Brain Thinking

Imagine walking into the Vertical Assembly Building at Cape Kennedy and seeing a hundred million pieces of machinery, equipment, devices, gadgets, and space junk scattered on the floor. There are nuts, bolts, titanium beams, silicon chips, solar cells, and four hundred and forty-five miles of wire. When reduced to this extent, you can't perceive these pieces combining together to create an earth-orbiting space station, it's all just space stuff to you. If you asked me to assemble it, I wouldn't have a clue, even with a blueprint, a set of tools, and lots of time on my hands. And I worked on the stupid thing. Right-brained thinking, assembly, is a lot more difficult than left-brained thinking, dispersion. It's easier to break things apart than to put them back together. It's easier to perceive the tree than it is the forest.

This doesn't mean that you're not a right-brained thinker. You are. We all have a right brain, and for most of us, it works quite well. You perceive a cloud in the sky as a white puffy thing, not condensed water droplets, not billions of suspended frozen crystals, not the pieces that make it up. You perceive the whole, making shapes out of it like a dragon from *Lord of the Rings* or Marilyn Monroe from *Some Like It Hot*. Watching your television, you perceive images, not little dots, not the 414,000 pixels that glow red, green, or blue, not the pieces that make up the screen. Your mind sees the picture holistically, using the right half of your brain. So, it's not that you don't know how to use your right brain, you do. It's just that you choose not to, you're taught not to; it's more natural for you to think deeply with the left. The right brain, it's believed, is for sunsets and rainbows. However, creative thinking needs right-brained skills, being able to see how the pieces fit together to solve the problem, the form that they make. Without this ability you get lost in the details.

While Walt Disney was a master of left-brained minutiae, he balanced this brilliantly with right-brained holistic thinking. He perceived Disneyland, at the highest level, as a movie starring his customers, his employees as cast members, and his buildings as movie sets. Disneyland was created with the use of storyboards, allowing Walt to conceive his designs from a right-brained-holistic point of view. This allowed him to see how the pieces would fit together, the overall impression they'd make, before he ever broke ground in the orange groves of Anaheim. Walt had a perfect sense of the whole and the parts, a balance between his right and left brain.

Right-brain thinking taken to an extreme, without the balance of the left, is impractical, flighty, and ineffectual. It's for sunsets and rainbows. Pure right-brained thinkers are philosophical, idealistic, and theoretical, unable to apply their thinking in any practical way. Right brainers can see the forest, but bump into the trees as they walk. The creative thinker, the effective thinker, like Disney, balances both right and left, and can perceive the parts and how they fit together to make the whole. Creative thinkers are whole brainers and the way they think, they way they create, is a process called self-organization. It's a process that uses both right- and left-minded skills.

Let me explain.

Whole-Brain Thinking and Self-Organization

Nature is the great architect, finding perfect harmony through a process of evolutionary design. Rain falls, self-organizes into streams, carves out magnificent canyons, flows to the sea, evaporates, forms into clouds, and then begins the journey as rain once more. Each time, carving the canyon a little deeper. The canyon is not the design of a magnificent executive, it happens naturally, through a process that creates itself. The world, as we know it, formed on its own. It self-organized. It's still doing so. You can see it happening every day.

Ross Ashby, a British engineer, coined the term *self-organizing system* in 1947 and described it as one with no executive function, no plans or goals; it's a process in which the components interact and find a balance through trial and error. For example, a flock of birds forms an overall pattern, a V-shaped design. But it's the interaction of the individual birds, each one flying behind and off center to the one in front, in an effort to gain aerodynamic lift from the other bird, that forms the V shape. It's not the result of a leader bird with a desire to create a giant V. In contrast, a high school marching band forms the word FIGHT on the football field by the leader's designating a place to stand for each band member. The drummer stands on the forty-yard line, next to the right hash mark, making part of the *F* while the trumpet player stands on the fifty-yard line, on the left hash mark, making part of the *H*. The birds are a self-organizing system, while the band is an executively formed one. Most of us think of design as being executively organized, like a marching band, or by the hand of an architect, or in the way God works; and not self-organized, like a flock of birds, like a flowing stream, or the way Nature works.

Ideally, nature is the better architect, creating designs in perfect balance, the result of a self-organizing process. The human mind, it-self the result of this process, does the same thing as it creates its own ideas. It works best when it lets the idea components interact and find a perfect harmony of pieces. This means using right-brain skills to perceive the whole and left-brained skills to see the pieces. The pieces provide positive and negative feedback, finding a balance between them, reconstructing, naturally, an idea that solves the problem made out of these borrowed pieces. Creative thinking is a whole-brained thinking process. Natural. The pieces fit together by interacting and forming a whole the way each bird in a flock knows to stay behind and slightly off-center of the one in front to form a giant V in the sky.

Of course, that's the theory. The steps in this book are designed to make it happen. Put simply, you don't know what your idea looks like, its form, its shape. You need to help it form, using borrowed ideas, combining them, making positive and negative judgments, exploiting the positives and eliminating the negatives, putting together new com-binations, redefining, replacing, and restructuring until it takes the form of a truly creative solution. This involves using your left brain to take your ideas apart and your right brain to put them back together. That's what this chapter is about, enhancing your ideas through the progression of self-organization. A genius does this naturally, in the shadows of the subconscious. It's a gift—the way a stream flows or a thunderstorm forms. You and I, however, don't possess this gift. Instead, we have to simulate it consciously. We have to step in and manually make it happen, force the interaction of ideas instead of just letting them come to pass in the shadows the way a genius does.

Let me explain.

The Simulation of Genius

According to Kurt Vonnegut, author of *Slaughterhouse-Five*, there are two kinds of writers: swoopers and bashers. Swoopers, he says, write quickly, higgledy-piggledy, not concerned about grammar, structure, or even getting the story right. It doesn't matter to them, because they come back and go over it, painstakingly, rewriting, restructuring, and replacing things that don't work. Bashers, on the other hand, create one sentence at a time, getting each exactly right before moving on to the next. When they're done, he says, they're done. No rewrite, no restructure, and no replacement. A basher is a genius. He can sense his entire work, holistically, even while he's focused on the detail of a sentence. He makes it all fit together as he creates it. Vonnegut was a basher. I'm a swooper; I have a hard time sensing my overall structure as I get lost in the detail of a sentence. The first drafts of this book were swooped, then painstakingly rewritten and restructured. Things that didn't work were replaced with things that do. Vonnegut was a genius. I'm not and never will be.

However, this doesn't mean that I can't create. What Vonnegut does subconsciously I have to do consciously. It takes me longer and requires the ability to edit, to see mistakes, and to enhance my ideas until they approach the tightly constructed creations of the genius. Genius does it naturally, letting ideas self-organize in the shadows of the mind, while I have to step in and consciously make corrections. I simulate the mind of a genius, even though I don't have the gift. You can too.

To imitate genius, the mind of a basher, slow down and alter your perception of your subject, look at a detail, then stop and remind yourself of how that detail fits into the overall structure of the subject. It's like climbing Whitney. You focus on your handhold, but every once in a while you look up, determine whether you're on route or not, making sure you don't get lost in the detail of the rock in front of you. Then adjust your pathway, retrace your steps if you have to, but always move forward, toward the top. Swoopers can make the summit like bashers, it just takes a little longer. Trial and error. And the ability to make the adjustments. Hell, F. Scott Fitzgerald was a swooper, so we're in good company. James Michener said of himself, "I'm not a very good writer, but I'm an excellent rewriter." This chapter, and the steps that follow, are swooper steps. They're designed to make corrections, for rewriting, and to enhance your ideas. Designed to simulate genius or borrow brilliance if you don't have it. As Ralph Waldo Emerson said, "Genius borrows nobly."

The Sixth Step as a Return to the Previous Steps

The sixth step is, quite literally, a return to the previous five steps. It's comprised of: re-defining the problem; re-borrowing the materials; re-combining the structure; re-incubating the solution; and re-judging it all again. You see, every time you return to one of these steps, you do so with more insight and a greater sense of creative intuition. As Figure 3 shows, you'll use judgment as the mechanism by which to drive the process in this step. In fact, judgment ultimately becomes the central part in the process.

FIGURE 3: THE SIXTH STEP

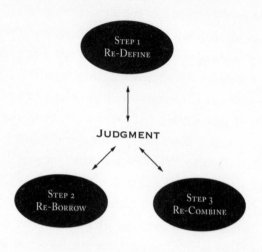

THE SIXTH STEP IS A RETURN TO THE PREVIOUS STEPS. THE INNOVATOR RE-DEFINES, RE-BORROWS, AND RE-COMBINES MATERIALS USING JUDGMENT TO DRIVE THE PROCESS.

Like climbing an unfamiliar mountain, this is a matter of trial and error. You'll construct a new idea using the first three steps, let it incubate, and then pass judgment on it. Using this judgment you'll identify the strengths and weaknesses of your solution and so then enhance the strengths and eliminate the weaknesses. Sometimes you'll have successfully defined the problem and never have to redefine it. Other times, you'll concentrate on the problem, struggling to grasp its true nature. Sometimes the components you borrow in the second step will fit together perfectly. Other times you'll have to return to this step and find more appropriate materials. Sometimes you'll have the perfect metaphor to structure your creation and never have to adjust it. Other times your metaphor won't extend as well and you'll spend the majority of your time adjusting it until it loses its original form. Each time you make an adjustment you'll have a new idea, the child of the one before it. Each adjustment, an enhancement or elimination, will be a step in the evolutionary process of the idea. Some adjustments will be minor, like subtracting a component to make the idea simpler. Others will be monumental, like using your creation to solve a completely different problem. Most will be somewhere in between the extremes. Sometimes you'll do this consciously, being well aware of each adjustment, other times you'll return to the fourth step and just let it all incubate again and so re-form in the shadows of your subconscious mind. As the idea evolves, you'll use your judgment of it to drive this process. For each project, the order you do these things in will change, the process will create itself the way nature creates a canyon. This is like the fight for survival in the evolution of a species. Judgment will tell you what flaws to eliminate, what strengths to enhance, and what step to return to next to achieve these things.

Re-defining the Problem

Everything in the creative process is based upon the definition of the problem. It's the foundation upon which all ideas are constructed. An ill-conceived problem leads to an ill-conceived solution. The problem determines the materials you gather to solve it. The problem structures the solution by combining things with a similar metaphor. And the problem determines the criteria upon which you judge and evolve the idea. Get it wrong and everything else you do is done in vain. So, you need to periodically return to the first step of *Borrowing Brilliance* and reconsider the problem. Are you solving the right one?

You can *re-define* the problem in varying degrees: in simple fashion—restating it, for example—or in a complex way, such as deciding to solve a different problem altogether. A creative thinker is a flexible thinker and so not afraid to rebuild the foundation. Whether you restate or completely change the problem, you need to be aware of your options—that it's not written in stone, that your foundation can be moved if it's built on an unstable surface. Over time, the problem will change. Sometimes the change will be subtle, other times it will be drastic.

Rewording, rephrasing, and restating a problem works to initiate new ideas because how you define a problem will determine how you solve it. Define it differently, even a little bit, and you'll solve it differently.

Henry Ford solved a very specific problem with the Model T that he described as "producing the least expensive car in the world." William Durant, CEO of General Motors, on the other hand, paraphrased this problem and *re-defined* it as "a car that people could afford." The difference? Henry Ford produced an inexpensive car. General Motors produced more interesting ones and more expensive ones, but solved the problem by establishing GMAC in 1919, a finance division, so people could "afford" them by paying monthly. Using this idea, and others, General Motors would eventually outsell Ford.

Often grammar is all you need to alter your perception of a problem. So use literary techniques to reword the problem, for example, experiment with positive and negative verbs. You could state your problem positively: *How can I improve the efficiency of our workers?* Or state the same problem negatively: *How can I reduce the time spent working on the product?* The positive, in this case, will focus on the worker and "how" she does her job. It will lead to solutions like better lighting or different tools. The negative will focus on tasks and lead to solutions like robotics or breaking the tasks into more manageable chunks. It's the same essential problem, but changing the point of view from positive to negative can change the solution it produces. You can also use active and passive statements. For example, you could use an active problem: *My workers are waiting for the next car to come down the assembly line.* Or a passive one: *The production line was slowed down.* The active statement focuses on the action, in this case, "waiting," and will produce ideas that will either eliminate the waiting (like speeding up the assembly line) or ideas that give the workers other things to do while waiting (like cleaning their tools). The passive statement will focus on the object, in this case the assembly line, and lead to ideas that solve assembly line problems (like rearranging the order in which things are assembled).

The next degree of *re-defining* is to choose a different problem from your matrix or hierarchy of problems. This is more than just restating the problem. It's choosing a different, albeit related, problem. Remember, CIA analyst Morgan Jones said the most common mistakes in problem solving are mistakes of scope, solving too high- or too low-level a problem. To combat this, you constructed a hierarchy and then made a decision as to which problem to solve within this hierarchy. So, you need to return periodically to it and reconsider the other problems in the matrix.

For example, Larry and Sergey are constantly going back to their matrix of problems and choosing new ones to solve. They go high and low. Nearly ten years after they developed their PageRank algorithm and founded Google, they announced that they were going to scan all the books of the New York Public Library as well as the libraries of Harvard, Stanford, and several other universities. Ironically, this was a return to the original problem Larry was assigned while working on his Ph.D. This lower-level problem had led them to the higher-level one of Internet search. In a press release, Larry said, "Even before we started Google, we dreamed of making the incredible breadth of information that librarians so lovingly organize searchable online." He added, "Google's mission is to organize the world's information, and we're excited to be working with libraries to help make this mission a reality." In other words, he's defining both a low-level problem and the way it relates to the overall hierarchy of problems. Google's highest-level problem, according to Larry at that time, was to organize the world's information. Google continues to innovate by solving problems throughout the entire hierarchy of problems.

But they didn't stop there. They continue to expand the hierarchy of problems and reach even higher. In 2004, Larry and Sergey did an interview with *Playboy* magazine just before Google went public. During the interview Sergey said, "Ultimately you want to have the entire world's knowledge connected directly to your mind."

"Is that what we have to look forward to?" the *Playboy* interviewer asked.

"To get closer to that—as close as possible," replied Sergey. "The smarter we can make the search engine, the better. Where will it lead? Who knows? But it's credible to imagine a leap as great as that from hunting through library stacks to a Google session, when we leap from today's search engines to having the entirety of the world's information as just one of our thoughts."

What you're seeing here is the expansion of a problem matrix. A few months later, Larry joined in and said at a Stanford lecture, "The ultimate search engine is something as smart as people—or smarter . . . For us, working on search is a way to work on artificial intelligence." Or put another way, in a different interview, he said, "On the more exciting front, you can imagine your brain being augmented by Google. For example, you think about something and your cell phone could whisper the answer into your ear."

At some point, though, you will be at a level of redefinition such that your problem breaks out of the problem matrix and you begin to solve completely different problems with your ideas. In the book *The Google Story*, author David Vise quotes Larry as saying, "Why not improve the brain? Perhaps in the future, we can attach a little version of Google that you just plug into your brain." This is clearly venturing into new territory, using the solution to Internet search to solve the problems of human intelligence.

Whether they succeed remains to be seen. However, using an existing idea to solve a completely different problem is the ultimate way of redefining your problem and is one of the most successful ways for thinkers to generate magnificent creations. An idea in search of a problem may seem counterintuitive, but it's very common in the history of innovation. For example, the history of the phonograph, which became the history of the Walkman and iPod, is the history of an idea seeking out a problem to solve.

The story begins in 1877 in the Menlo Park, New Jersey, laboratory of Thomas Edison. The Wizard of Menlo Park, as he was popularly known, was working on an automated relay device for a telegraph station. At the time, a telegraph signal could travel forty miles or so, over the crude wiring system strung up along the nation's railroad tracks, before it would lose its strength. So relay stations were established every forty miles. They were manned by an operator who read the incoming sequence of dots and dashes, recorded them by hand on paper, and then turned around and repeated them to the next station down the line. Edison wanted to automate this, record the dots and dashes on ticker tape, and then feed this tape into another machine that reproduced the message and sent it down the line. This eliminated the operator and operator error. While experimenting with a prototype, he noticed the dots and dashes, when replayed at a high speed, sounded like human voices. This intrigued him. So Edison abandoned his original problem, redefined it, and started working on the problem of reproducing human voices, using the equipment he had invented for a different purpose.

What began as a telegraph relay problem turned into a voice recording problem. He sketched a machine with a diaphragm which had an embossing point that was held against a ticker tape as it moved through the machine (similar to a telegraph design). In the next iteration, he replaced the ticker tape with a metal cylinder wrapped in tinfoil. He added another diaphragm for playback. Theoretically, it would work like this: A person would speak into a mouthpiece, the sound waves would move the diaphragm, which would move the needle, which made indentations on the metal cylinder. For playback, the process would be reversed. The metal cylinder with the recorded indentations would be rotated, moving the needle, which moved the diaphragm, which turned the incised patterns back into sound waves. Edison gave these sketches to his mechanic, John Kruesi, and as legend has it, Kruesi built it in less than two days while Edison hovered over his shoulder watching and making construction suggestions. When they were done, Edison leaned over the machine while Kruesi hand-cranked the metal cylinder. Edison spoke into the mouthpiece the children's nursery rhyme, "Mary had a little lamb, its fleece was white as snow . . ." When Edison was done, Kruesi reset the needle on the cylinder and cranked it again at the same speed as when Edison spoke. To their complete astonishment, the machine repeated, "Mary had a little lamb . . ." in Edison's tinny voice. It worked on the very first try! Thomas Edison had successfully recorded the human voice for the first time and was able to replay it. The phonograph was born, the child of the telegraph, thanks to the flexibility of its inventor to change completely the problem he was working on.

On January 24, 1878, the Edison Speaking Phonograph Company was established to manufacture and sell the new invention. Edison was given ten thousand dollars for the manufacturing rights and a promise of 20 percent of all of the profits. However, no profits came. You see, the phonograph, a result of solving a telegraph problem, continued to be an idea in search of a problem to solve. In an 1878 magazine interview Edison explained what people could use it for: a) letter writing; b) phonographic books for the blind; c) teaching elocution (how to pronounce words); d) family records (to record the dying words of people); e) speaking toys; f) clocks to announce the time; g) classroom instruction (to replace teachers); h) preservation of language; i) recording telephone conversations; and as j) music boxes. As a novelty the Edison Speaking Machine was a hit. As a practical device it was a complete failure, it couldn't solve any of those problems, and it would remain a failure for the next ten years. It was a good idea in search of a well-defined problem to solve. Ultimately, it would revolutionize the way people entertained themselves and give birth to the Walkman and iPod a hundred years later.

So, as you enter the first stage of the enhancement step, stop to reconsider the problem. Ask yourself: *Can this idea be used to solve a different problem?* There are numerous ways of answering this question. You can simply restate the problem you're working on the way Durant restated Ford's problem. You can choose another problem in your overall hierarchy of problems, the way Larry and Sergey do. Or you can completely change the problem you're working on, the way that Thomas Edison used a telegraph solution to solve a phonographic problem. Instead of selling your product to men to shave their faces, maybe you redefine its purpose, and sell it to women to shave their legs. Either way, never lose awareness of the problem, and never etch it in stone. This isn't the SAT or a college midterm. You have control over the problem, so don't be afraid to exercise it. The evolution of an idea includes the evolution of the problem it's solving. Redefine your problem. It's the hallmark of a creative mind.

Re-borrowing the Materials

Constructing a new idea is more like solving a jigsaw puzzle than it is like building the Opera House on the shores of Sydney Harbor. There are no blueprints, it's a haphazard process defined by trial and error. In the second step of *Borrowing Brilliance*, you looked for places with a similar problem. You started in your own domain, taking pieces of ideas from your competitors. Then you took a step away and borrowed ideas from other industries that were solving a similar problem. Then, if you're good, you went outside the domain of business and found places in science, for example, or entertainment that had a similar problem and borrowed puzzle pieces from there. Then you took this stuff and began to combine it by finding an appropriate metaphor. However, the metaphor is never perfect and only begins to structure a solution. You discovered the problems with the metaphor when you put it under pressure by passing judgment on it. The pressure, undoubtedly, created new problems, which every solution does. Now you need to begin fixing it by using more appropriate materials. In other words, replacing ill-fitting components with better-fitting ones. As you begin to solve a jigsaw puzzle, open sections of it will show you the form of the piece that you're looking for. *Re-borrowing* materials is that search for better stuff to better solve the problem.

Put simply, *re-borrowing* is really replacing components in an existing structure, inserting new, substitute material into it. For example, in 1452, Johannes Gutenberg constructed the modern printing press by using xylography as the overarching metaphor for his creation. Xylography is better known as woodblock printing and had been around for hundreds of years by the time that Gutenberg began to study it. Using a wooden block, a printer painstakingly whittled a page into it, lathered it with an egg-based tempera ink, and then printed onto pieces of vellum or calfskin by pressing the woodblock onto the vellum by hand. It was a time-consuming process, sometimes taking years to produce a single book.

Some historians and scholars agree that the invention of the Gutenberg printing press was the most important invention in the history of mankind. It allowed for the great dissemination of information and the education of the masses. Before Gutenberg, education was verbal and books were scarce and only available to the elite upper class. After him, books were inexpensive, education became common and accessible, and the incredibly creative period known as the Enlightenment occurred, because now ideas could be passed along more easily and combined with other ideas to form new ones. Gutenberg provided the materials that Isaac Newton and those who followed needed in order to receive and pass along their ideas.

The evolution of the printing press used replacement, the re-borrowing of materials, as the primary methodology for construction. The wood-block printing process provided the structural metaphor. Gutenberg merely replaced the xylographic components with different ones. Instead of using a woodblock, Gutenberg replaced it with metallic letters attached to a metal plate. He *reborrowed* this idea from weapon and coin forging, which had been around for centuries. Then, instead of printing on vellum, he replaced it with the much cheaper material called papyrus or paper. Paper, like coin forging, had been around for centuries but wasn't used for books because it wasn't as durable as vellum. Then, instead of using the egg-based tempura, Gutenberg replaced it with a much cheaper oil-based ink. This ink smeared on the vellum, but when combined with paper it worked extremely well because paper absorbed the ink, making it smearproof. Finally, instead of hand-pressing the block, he replaced this with a screw press that he borrowed from winemakers and olive oil producers. All of these existing things were combined by replacing the components within an existing overall structure, in this case the metaphor of wood-block printing, and as a result one of the most important inventions in the history of the world was created.

Re-borrowing or replacement isn't unique to the printing press. It's part of the evolutionary process of all creations. New components replace existing ones, making a new idea even though the overall structure hasn't changed. Gutenberg went out and searched for new materials that could replace the existing ones. He used his understanding of the problems to find these new things. Instead of using a wood-block, he asked himself, what other material could he use to shape into letters, words, sentences, and paragraphs? Well, he answered, coins use metal to form letters, numbers, and other shapes. Maybe I can use metal, too, he said. Instead of printing on vellum, he asked himself, what other material could I print on that might not be so expensive? It was these series of questions, or ones like them, that led him to his invention. First he borrowed from within his domain, using xylography as his overarching metaphor. Then he borrowed from a different industry, the winemaking and olive oil making domains. Finally, he ventured even farther away and borrowed from the military and government, who were using metallurgy to create weapons and coins.

Similarly, Edison's phonograph evolved when others replaced components in the original structure that he'd created. Edison's machine, as I said, was an interesting novelty but it didn't work very well. So, it was put on the back burner because over the next decade there was a burst of creative inventions that silenced the phonograph. Fast-shutter motion photography was invented and evolving into motion pictures. Edison invented a working electric light bulb and was beginning to install electrical power plants. And Alexander Graham Bell invented the telephone.

In 1887, Charles Tainter and Chichester Bell released a new machine they called a "graphophone." Using Edison's machine as the overall structure, they replaced the tinfoil drum with a cylinder made entirely of wax. This allowed for longer playing time and more clearly defined recordings. Instead of using a fixed needle, like Edison, they replaced it with a floating stylus that produced a cleaner sound and resolved the pitch fluctuation problems Edison's machine had. Finally, they replaced the hand crank with an electric motor, which allowed for consistent recording and playback speeds. The graphophone was a huge improvement over Edison's phonograph. Units were installed in "entertainment parlors" and used as coin-operated vending machines to play music alongside other machines called kinetoscopes that were used to play movies. This renewed Edison's interest, but he felt his machine should have the more dignified purpose of being a dictating machine, not a coin-operated amusement. He went off in a different direction trying to solve a different problem, leaving the fate of the phonograph in the hands of others.

Ignoring Edison, the others continued to work on the entertainment problem. A small recording industry sprang up and began to sell prerecorded cylinders. However, this was time-consuming, since each cylinder required a new performance. Then, in 1893, Emile Berliner invented the "gramophone," which was similar to the graphophone except that Berliner replaced the cylinder with a hard rubber disc. The discs were much cheaper to produce and thousands of copies could be made from a zinc master. This solved the problem of a new performance for every cylinder, since an artist could perform once onto a zinc disk that could then be repeatedly copied. The music industry was born. A modern compact disc (CD) is copied from what's called a "gold master," a descendant of Berliner's zinc master. Eventually, the rubber discs were replaced by vinyl ones and called "records" and the gramophone was renamed the phonograph, the original name of the machine from which the new device had evolved. We could continue to follow this evolution into the Walkman and iPod, but I'm sure you get the point. Things can evolve by the replacement of existing components with new ones within an existing structure. It's simply a return to the second step, and this is why I call it reborrowing.

Of course, it's not just mechanical things that work this way. Conceptual things do too. Several days before he was to take the oath of office as president of the United States, John F. Kennedy worked with Ted Sorensen, his chief speechwriter, to construct a powerful inaugural address. Borrowing from literature, sermons, the Declaration of Independence, and historical speeches, Kennedy and Sorensen carefully crafted the address. However, Kennedy knew that he needed a "hook," a clever statement that would be controversial, fitting, and memorable. He tried different things but none seemed to work. Then he remembered the headmaster at Choate prep school, the Connecticut boarding school he attended as a teenager. You see, the headmaster at Choate was known for telling his students, "Ask not what Choate can do for you, but what you can do for Choate." Kennedy brilliantly borrowed this line, replacing one component for another, and said, "Ask not what your country can do for you, ask what you can do for your country." It's still considered one of the greatest speeches in American history.

So, as you work to enhance your idea, study it and ask yourself: *What components can I replace in this structure to make it more effective?* This is a return to the second step, except now you're smarter when you return. Your judgment has defined what you're looking for, it has established your creative intuition—an idea you can describe as "having these strengths" without having "those weaknesses." So, you simply ask yourself where to find this component you've described. George Lucas used Joseph Campbell's *monomyth* as the plot, or structure, for his movie and replaced the components with science fiction ones. Charles Darwin took the structure of Charles Lyell's theory of geological evolution and replaced its components with biological ones. And I can take an existing direct marketing structure—the offer, timing, package, and list—and replace it with my own components to make a new mail program. Replacing components, reborrowing, doesn't alter the overall metaphorical structure, it only makes it more effective, since no metaphor is perfect. However, you can, and should, alter the metaphor itself for the same reason.

Re-combining the Structure

Remember what Sigmund Freud said: "We have constantly to keep changing these analogies, for none of them lasts us long enough." In the beginning, the metaphor provides the structure for your idea, but it's far from a flawless one. Metaphors are never perfect. This will become apparent once you start extending your own metaphor and once it has passed through your negative viewpoint. The solution will always present problems. It won't fit just right. The evolution of an idea solves these subordinate problems by using the replacement of components, as you just saw, as well as the *recombining* of the metaphor, a return to the third step of *Borrowing Brilliance*. That's why, in part, metaphors are lost in Emerson's literal graveyard. We don't perceive the horse-drawn-carriage metaphor in the modern automobile; the structure has altered so much that the literal carriage has receded deep into the DNA of the machine, until by now the extant remnant of it revives only when we speak of "horsepower." Nonetheless, the automobile is a descendent of the carriage.

Like reborrowing, there are varying degrees to *recombining*. They can be as simple as eliminating a useless component or as complex as completely rearranging the components to form an entirely new structure. While metaphors are critical to gaining insight into a new creation, don't be seduced by them, don't be afraid to alter them to better suit your unique situation. That's what returning to the combination step is all about. In every creative thinking process you'll spend some of your time altering the structure. While there are an infinite number of ways to *recombine*, I use five thinking tools: addition, subtraction, multiplication, division, and rearrangement. In other words, to restructure an idea I add new components, subtract others, multiply components, divide the idea by splitting a group of components off to stand alone, or simply rearrange the existing components in a new order. All of this is done in an effort to evolve my idea by eliminating weaknesses and enhancing the strengths.

In the first chapter, you learned that every solution creates a new set of problems. As your idea evolves, it will undoubtedly create problems you didn't anticipate. So things are added to solve these new problems and your idea becomes more complex over time. You can see this in the evolution of any machine, product, or theory. Today's Boeing 787 has over six million different parts as compared to a few hundred contained in the original Wright Flyer that flew at Kitty Hawk a hundred years ago. Darwin's theory of evolution by natural selection was presented in a book with fewer than five hundred pages. Today, evolutionary biology has been combined with genetics and hundreds of other theories and concepts, producing a domain so complex and vast that it would be impossible for anyone to comprehend it all. People in evolutionary biology have to specialize. Every solution creates new problems and so new things are added to solve these problems. This is the way things evolve.

Edison's original phonograph was a brilliant innovation, but it created a new set of problems that made it essentially useless. First, the tinfoil cylinders were so fragile that they would only last for a few replays. This problem alone hampered the use of the machine, making it a novelty. So, things were added. Instead of a fixed needle that tended to gouge out the tinfoil, a floating stylus was added that prevented the destruction of the cylinder upon replay. However, this increased the complexity. The stylus required a number of working parts, like tiny springs, that the fixed needle didn't. Once Tainter and Bell replaced the cylinder with a rubber disc and electric motor, making the machine practical, this created dozens of other new parts—turntables, switches, wiring, and complex gears. Just look at a photograph of Edison's talking machine as compared to the Tainter/Bell graphophone and you can see the complexity that addition had created. Today's iPod, while much smaller, is so complex that the tiny silicon chips are understood by few, including the designers of the iPod itself. Thousands of miniature things are contained in your iPod. Don't be misled by the simplicity of its design—though the descendant of Edison's simple talking machine, it's a very complex device.

As you identify new problems that your solution has created, ask yourself: *What components can I add to my idea that will solve these additional problems?* Use restraint, don't overcomplicate your idea. In fact, as your idea evolves and things are added, make sure you stop and consider subtracting things too.

Like adding, subtracting changes the form of a structure, obscuring the original metaphor. Unlike adding, it simplifies instead of complicates. For this reason it's a tool I often use, like the putter in my golf bag—I like to finish the combining hole with this club. It's a way to clean up your ideas, streamlining them, focusing and making them easier to understand and so more useful.

Both intellectual and biological evolution are naturally additive, elements put in to solve problems and mutations formed that are used to fight for survival. This causes each to end up with legacy components. In biology, these are called vestigial organs, like the small vestigial leg bones buried deep in the back of a modern blue whale, left over from the days when its ancestors walked on land. In the realm of ideas, I call them vestigial components, for example a screen saver on a laptop computer, left over from the time when desktop computers used cathode ray tubes and required screen savers to prevent an image from burning into the tube if it was left on the screen for too long. It's no longer needed but still there. Look for these vestigial things in your ideas and use subtraction to eliminate them.

When Masaru Ibuka combined the idea of a miniature tape recorder with a lightweight pair of headphones, he quickly realized he had a vestigial component. The machine no longer needed a recording mechanism because it was solving a different problem—playback. So he instructed his engineers to subtract it, making the product even smaller, cleaner, and more elegant. However, this seemed a bizarre request to the employees of the Tape "Recorder" Division of the Sony Corporation.

As you begin the enhancing process, the first order of business is to clean up the idea. Simplify it. Look for things that you added in the construction process that, in retrospect, aren't aiding in the solution. Using this tool, identify them and eliminate them. Einstein said, "Make everything as simple as possible."

My boss at McDonnell Douglas, Bob Pedralia, was the master of simplicity in an aerospace world defined by complexity. He used simple words, never used acronyms, and liked to restate complicated concepts using clean, austere explanations. An engineer would tell him, "The launch vehicle vector created by the supplementary payload is greater than the compensating vector we can accomplish with the forward discharging thrust." Bob would look at him, puzzled, and then say, "Do you mean it's too heavy?" And the engineer would stop, think for a moment, and then confess, "Yes." It was brilliance in action.

A sculptor chips away at her block of marble. She enhances it by eliminating things. She simplifies it, removing the imperfections from her design. And so it is with any idea, any conception. Use subtraction to enhance the idea, removing the imperfections, negative things, and unnecessary things. It's a matter of taking away the inessential and so revealing the fundamental nature of your solution.

In business, simplicity is critical because the competition for solutions is fierce. According to marketing expert Al Ries, every day the average consumer is confronted with thousands of different selling messages in the form of television ads, print ads, banners on the Internet, billboards, salespeople, radio commercials, and even signs above the urinal in the men's room. Good luck trying to get your message through all this clutter, especially if what you're selling is complicated. You need a simple, focused, singular idea if you want to succeed. You want to make ideas that survive, memes that get passed from person to person, successfully competing against the other memes. So you have to provide the world with a simple idea, something easy to understand and pass, or else it gets lost in the shuffle. Unnecessary complexity kills it.

To employ this tool, ask: *What can I take away that's of no use?* Or you might ask: *What component is no longer needed?* Or even: *How can I focus the intent of my idea by eliminating useless or vestigial components?* Or just, *What can I subtract?* Or, *How can I simplify?* The sculptor begins her creation with a block of marble and then subtracts to create. This is the same way I like to construct my ideas. I solve as many problems as I can, building a conceptual block of marble, and then I use this tool to trim it out and complete my creation. It's how I wrote this book. First I constructed two thousand pages of manuscript and then I subtracted eighteen hundred, simplifying it, to get to the two hundred you hold.

A warning, though—the Einstein quote above was not complete. What he actually said was, "Make everything as simple as possible, but not simpler." In other words, simplify to the point where you just have the crucial elements but not to the point where you start removing them as well. I could simplify the design of tax software by removing the on-screen help function, but it's crucial to completing a tax return and so shouldn't be eliminated. At the end of the day, complexity wins the game. It's the nature of things.

The next tool, idea multiplication, is like idea addition except that instead of adding a new component you add an existing one over and over, repeating it, usually as the primary problem-solving element. For example, the first powered aircraft, the Wright Flyer, was actually a biplane, its primary element multiplied by two, providing stability and rigidity for the structurally weak craft as well as additional lift. A few years later, Manfred Albrecht Freiherr von Richthofen multiplied the primary wing by three and created a triplane that was more maneuverable than the biplane and so enabled him to control the skies over the European battlefields. They called him the Red Baron.

In 1971, Gillette introduced a double-blade razor called the Trac II. The first blade, they claimed, pulled at the whisker and the second blade chopped it off. It was a huge success. But for some reason twenty-seven years passed before Gillette multiplied the blade by three and released the Mach 3. Then they got smart. A few years later they multiplied by five and created the Fusion. (For some reason they skipped four blades. *Go figure, right?*)

In 1967 the Beach Boys released one of the most influential albums in rock and roll history. Produced by Brian Wilson, the group's founder, *Pet Sounds* was a production masterpiece as much as it was a musical one. Using the recording equipment, Wilson multiplied his voice, combining two identical tracks, staggering them slightly, and creating a unique harmony with himself. This multiplication technique is still used in pop music today. Listen closely to Britney Spears—unfortunately, there's two of her singing.

To use this tool, ask: *What component can I multiply to more effectively solve my problem?* In most cases it will be the primary component of your design, like a voice in a song, a blade in a razor, or a wing in an airplane. For example, Squaw Valley Ski Area, at Lake Tahoe, has a cable car called a funitel, which can run in near hurricane conditions because it's suspended by two cables instead of one and so eliminates the fulcrum swinging effect of a normal cable car in a high wind.

It follows if multiplication is like addition, then division is like subtraction. The difference is instead of eliminating a single component, a group of components are split off and become a stand-alone idea. Subtracting simplifies an existing idea, while division creates a new idea out of the old one. For example, Edison's original phonograph had two primary functions: It could record and play sound. However, the more successful graphophone was actually a division of the phonograph because it could only replay prerecorded cylinders. It split the original intent of Edison's machine and focused on just one problem and the components needed to solve it. As you know, it was much more successful.

Within hours of conception, a human fetus begins growing by dividing itself. A single cell splits in two, in a process called mitosis, and as it continues to grow it experiences cell division over and over. In fact, you'll experience about ten thousand trillion cell divisions in your lifetime (if you're lucky). Some plants reproduce this way, in a process called asexual reproduction (they're unlucky), forming entirely new organisms simply by dividing themselves. Your thoughts can form the same way: a new thought dividing, splitting off from an original thought and becoming, of itself, a fresh idea that stands on its own merits.

Thought division is how product categories are created in business. Marketers and entrepreneurs isolate a unique use of an existing product, a unique problem being solved, and then create a new product to serve this isolated use. They divide products, spin them off, to create new ones and new categories. For example, automobiles were originally hobbyist's vehicles used by wealthy playboys. But other customers began to use them to haul materials. So designers created a new category of automobile by dividing its function. Now they had cars and trucks. Cars were then divided between luxury and family cars. Trucks divided into pickup trucks, vans, and delivery trucks. Today there are dozens of automobile categories—sports cars, minivans, SUVs, hybrids, compacts, and subcompacts—each the evolutionary descendant of another category, usually the result of designers' dividing the use of the product. While each product is easier to use, simpler for the target customer, the overall category—automobiles—becomes more and more complex. Just another example of how creativity complicates the world.

Entrepreneurs or would-be entrepreneurs take note—this is one of the best tools you can use to create a new business. For example, a small company, TechSmith, has created a popular software product called Snagit. The product is used to take digital pictures of your computer screen the way the "print screen" button works on any computer. In other words, the creators simply divided the functions of a computer, separating this feature and making a product out of it. They have taken an old idea and created a new product out of it.

To use division, ask: *Is there a part of this existing idea that I can separate and create something that will stand alone to solve another problem?* The original *Star Wars* script was more than three hundred pages long and incredibly complex, so Lucas divided it into three stand-alone movies (known as the Trilogy.) Sometimes division works better than subtraction. When combined with the technique of redefining the problem it's a powerful thinking tool.

Finally, your last restructuring device is the rearrangement tool. An idea is a pattern created by the material components, their volume, and how that material is connected. A pattern is what I've been calling an idea's structure. You can create a new pattern, a new structure, by simply rearranging the existing material of an idea. There are seven notes in music, yet they can be arranged in an infinite number of ways to create an infinite number of unique melodies. In fact, composing music and arranging it are sometimes called identical skills, for arranging is at the heart of writing music, or writing anything for that matter. To write this book I had to arrange four hundred thousand letters in a unique order. They're from the same twenty-six-letter alphabet that Hemingway used to write *The Old Man and the Sea* and Jefferson used to construct the Declaration of Independence, except I arranged them differently, creating a different pattern and so a different idea. As Nathaniel Hawthorne said, "Words . . . how potent for good and evil they become in the hands of one who knows how to combine them."

As a Web designer you can rearrange the elements of a screen in order to make it more effective. You can watch how users interact with the site and place the most popular elements or the ones that will drive sales in the area that the eye is drawn to (which, according to design expert Don Norman, is in the upper left-hand part of the screen). You can arrange components of the site in a way that's convenient for keystrokes or mouse movements, to make it more user-friendly. You aren't changing the content, just rearranging it in such a way as to make the site more effective. It's like rearranging the furniture in your living room. Putting the couch in front of the bay window creates a different feel to the room than having it up against the back wall. Same stuff in the room, but a different sense is created. The Chinese call this feng shui—the practice of arranging objects, like furniture, to find a perfect balance between them and the natural order of things.

Rearranging the parts is simple but can make the difference between a good idea and a great one. For example, an e-mail campaign, a series of contacts to potential customers, can be restructured by changing the order in which they are sent; instead of sending the discount offer before the newsletter, do the opposite, send the newsletter and then the discount—your effort might work better that way. When Jonathan Ive designed the first iPod he restructured the components—the screen, play button, pause, forward, back, and on/off switch—into different configurations, putting the play button on top in one try, then restructuring it and putting it on the bottom in the next try. Remember, Albert Einstein defined creativity as "combinatory" play. And that's exactly what the rearrangement tool is, it's combinatory play.

Often, rearrangement takes the form of reversal. For example, in 1820, scientist Hans Christian Oersted, experimenting with wires, batteries, and magnets, noticed that a compass needle moved when a wire carrying current was passed over it. *Hmm. Interesting observation,* he thought. Several years later, Michael Faraday, while playing his own combinatory games, wondered what would happen if he reversed this experiment and passed the magnet over the wire. Much to his delight he created an electrical current in the wire by rotating the magnet around it. The faster he spun the magnet, the greater the current. Today, this is how we generate electricity in a power plant. In a hydroelectric plant, water is used to rotate magnets in a turbine that spins around a wire and creates electricity. In a nuclear power plant, water is boiled from the heat of fusion, which creates steam, which spins a magnetic turbine around a wire and creates electricity. By borrowing Oersted's experiment and reversing it, Faraday discovered something that changed the landscape of the world.

Henry Ford didn't invent the automobile or the assembly line. He did, however, create the modern "moving" assembly line by borrowing the idea from a meat packing company. The meat packers used a moving hook and conveyor system to disassemble a cow. Ford reversed this and used the same idea to assemble his automobiles. He simply rearranged an idea in a different business so that it worked to solve a problem in his business.

Quentin Tarantino created a bizarre effect in *Pulp Fiction* by reversing the order of the scenes. The movie begins with Travolta and Jackson, the main characters, in a café and then cuts to a scene from several days earlier, and then runs backward until the action reaches the café again.

To use this tool, note the pattern that your structure creates and then make adjustments to that pattern. Ask yourself: *How can I rearrange my idea to better solve my problem?* Think of it like Einstein did, as combinatory play. Change the order in which you contact your customer. Move elements around on your Web page. Or reverse things, the way Oprah did when she reversed the letters in her name to come up with a creative solution for the name of her production company, Harpo Productions. Without rearrangement, the lights in your home wouldn't work: Faraday would not have played around with the combinations of borrowed magnets, wires, and batteries with which he induced an electric current.

The structure of your idea can take on an infinite number of shapes: hierarchical, linear, circular, or some haphazard one with an unfamiliar form. Use addition, subtraction, multiplication, division, and rearrangement to change this shape to eliminate the weakness and enhance the strength of your idea. Restructure to see if this better solves your problem. Sometimes you'll need to start over with a completely different metaphor, the way George Lucas did as he struggled with the *Star Wars* script. Other times the adjustment of your metaphor, using these *re-combining* tools, will work to solve your problem and your metaphor will be lost. Of course, this is a process of trial and error, and as it evolves, like Lucas, you'll get stuck and not know what to do next. That's when it's time to reheat your idea.

Re-incubating the Solution

The enhancement of an idea is a complex endeavor. *Re-defining*, *re-borrowing*, and *re-combining* is as much an art as it is a science. Millions, perhaps billions, of combinations are possible with even the simplest idea. There's so much material out there to combine that it staggers the mind to think about it. And yet, you'll find yourself trapped in the natural thinking holes that the conscious mind creates. Your idea, whether it's good or bad, will begin to excavate a path in your mind, a path that you'll naturally follow and that consequently will keep leading you to the same conclusion. When you see this happening, you'll know that it's time to reheat your solution, it's time to put it away, to return to the fourth step, and fill in the thinking holes that you're digging for yourself. You see, enhancement can take place in the light of consciousness, or in the dim shadows of the subconscious mind.

Learning to be a creative thinker is no different from learning any subject or any endeavor. It begins with rote learning. At first you consciously perform tasks. If you're learning to play basketball, then you practice your set shot, your jump shot, your hook, and ball handling techniques. Through conscious repetition these things become second nature, relegated to your subconscious mind. You're no longer aware of how you release the ball on your jump shot, it happens naturally. The same is true with creative thinking. First you'll perform the creative tasks consciously: defining, borrowing, combining, incubating, and judging. Then they'll start to become more natural. You'll begin defining, borrowing, combining, in your subconscious mind. Even judgment will start to become more natural. Soon you'll adopt Steve Jobs's contradictory personality. You can help facilitate this process by forcing the incubation of ideas, by consciously putting your ideas away so that they can re-form in the shadows, away from the harsh light of conscious thought. Your subconscious is a powerful ally in the creative process and you need to invite it into the game.

Remember, there are three stages to subconscious thought: the input stage; the incubation stage; and the output stage. The input stage is the conscious defining, borrowing, combining, and judging. You present your subconscious mind with these things. Next, the incubation stage is giving the subconscious mind time to work on these things, time to *redefine, reborrow, recombine*, and *rejudge* the ideas you're working on. The interval can be a brief moment, a pause, or days, weeks, or even years of desertion. Finally, the output stage is the listening stage. It involves relaxing Freud's watchman, the gatekeeper to conscious thought. It also involves quieting the logical mind, clearing out conscious thought, to make room for the subconscious to speak. You can't listen to yourself if you're constantly talking to yourself.

So, as your idea evolves, use your subconscious to aid in the evolution. Force the materials into the shadows of your mind. Wait. Then quietly listen for a response. Make it a habit. Use the creative pause as you think and talk about your ideas. Simply stop for a moment or two, and listen for a new thought, a tangent that your shadow self may want to lead you on. Creative thinking is not necessarily efficient thinking. It's trial and error. Your mind will take you on paths that lead to nowhere. That's just the nature of the process. It's like climbing the Mountaineers Route on Whitney. You'll climb up, get trapped and so have to retrace your steps, and then start up in a different direction. Be prepared to abandon these paths. You don't know exactly where you're going and whether you're on route or not. The only way to tell is to try. And keep trying. Over time, you'll begin to teach your subconscious how to construct better ideas by first consciously doing it. Over time, the skills will be transferred to your shadow self and you'll get better and better at creative thinking.

Of course, you and I will never be a George Lucas, Isaac Newton, or Steve Jobs. They're gifted people who do this innately, it's already part of their subconscious makeup. But you can become a better thinker. You can be more creative. Begin to replicate the mind of a genius. And make your own contributions to whatever domain you choose to study.

It's fun stuff and can be very rewarding, but it takes a lot of work. The world is becoming more and more complex and ideas, products, and theories are being created at an alarming rate. Historically, access to information and education were a critical component to success. Now, since we all have access—the game has changed—it's the creation of information that determines success. You can't do this without the help of your shadow self. It's impossible.

But don't fret. If your ideas seem frivolous or like cheap imitations, that's okay. That's exactly how Lucas, Newton, and Jobs felt. They just kept working at it. They understood that innovation is an evolutionary process.

Re-judging It All

As the process evolves, you'll leave a stream of discarded ideas. Things that didn't work. Pathways that led to nowhere. At first blush, this may seem incredibly inefficient. But it isn't. Your failures are teaching you. They're fossils in the evolution of a new idea. As Henry Ford said, "Failure is simply the opportunity to begin again, this time more intelligently." The more ideas you have, the more judgment you pass, and so the greater your list of positive and negative attributes becomes. This accumulation of judgments further hones your creative intuition, making you more intelligent. Over time, you'll know exactly what you're looking for, and you'll know how to describe it, even if you don't know what it is. As I made my way up the Mountaineers Chute, I began to become smarter about where I placed my hands and feet. I realized that the darker rocks tended to be loose and so tried to find a vein of lightly colored ones to climb upon. The same is true as you search for your solution. You'll begin to recognize ill-fitting ideas and gravitate to the ones that have the positive attributes that you've accumulated.

As things unfold, periodically review your ideas using the judgment viewpoints that you acquired in the fifth chapter. Remember, judgment isn't the search for an ultimate reality, judgment is based upon a viewpoint. You can perceive anything from either a positive or negative point of view. It's just a "hat" you put on. So, when your subconscious mind serves you a new idea, take a step back, put on your positive hat, and ask: *What are the encouraging attributes of this idea?* Then list them. Remember to rely on your definition of the problem to provide you with the criteria for this judgment. Once you've done that, take another step back, put on your negative hat, and ask: *What are the harmful attributes of this idea?* Then list them as well. As you look at this list, one point of view will usually overshadow the other. Positive or negative will "win." You'll debate yourself, using logic and evidence to support your claims and determine the winner. Then you'll put all that away, clear out your mind, and ask: *How do I feel about this idea?* This is, as you have learned, a checkpoint to see if logic is in agreement with feeling and emotion. If it isn't, then you know to return to the problem. This usually means that you've defined it incorrectly, that you're using the wrong criteria to judge your ideas.

Think of yourself as a sharpshooter doing target practice. Each new idea is a shot at the solution. An accomplished marksman uses each miss to adjust his next shot. If his shot is just to the left of the bull's-eye, he moves the crosshairs slightly to the right on the next shot. He's using a failed attempt to enhance his next shot and get closer and closer to the center of the target. Using positive and negative judgment, the way a sniper uses his missed shot, over time you'll zero in on the solution.

Doing this is what Ford meant by starting over more intelligently. You'll return to the problem with deeper insight into it. Also, since you can now describe your perfect solution by calling it one with "these positive" attributes and without "those negative" attributes, you'll become a much more effective idea thief. You'll know, through creative intuition, where to look for an idea like that, and you'll be more apt to recognize it when you see it. Then you'll replace these loose-fitting components with better-fitting ones. Your designs will tighten up as you carefully put your jigsaw puzzle together. When you see something, you'll be like Steve Jobs jumping around the room at Xerox PARC screaming, "This is the greatest thing," and, "Why aren't you doing anything with this?" Because you'll know exactly how it fits into the intellectual jigsaw puzzle you've constructed in your mind. And when you return to the combinatory step, you'll be able to make adjustments. You'll eliminate those vestigial components that complicate your solution, while at the same time adding new components to help solve the additional problems that your solution creates. This is the evolution of an idea, and it's driven forward by the appropriate use of judgment, by periodic review of your solution.

Brilliantly Borrowed Brainstorming

Now, more than ever, creativity and innovation are the lifeblood of any organization. Product life-cycles have grown shorter and shorter. New ideas are being introduced into the marketplace at a feverish pace, unseating old ideas and leaving a path of corporate destruction in their wake. Innovation and creativity now drive the market, replacing scarcity and price as the primary keys to success. It's a wave that's just beginning to crest, and you'll need to ride that wave or else drown in the turbulence of its wake.

While creative thought, ultimately, takes place in the shadows of the mind, this doesn't mean you can't incorporate it into the processes of an organization. My time spent as an innovation leader at Intuit and other Fortune 500 companies has taught me that you can incorporate creative thinking into the DNA of any organization, by using the ideas in this book to alter the brainstorming process.

Most companies have either a formal or an informal brainstorming process, often mediated by an outside source. However, the great misconceptions that result from brainstorming are more detrimental to the creative process than anything that results from the process itself. The constraints of brainstorming leave out many important aspects of the creative thinking process. In particular, most brainstorming moderators tell you to suspend judgment, and most of us erroneously take this to mean that criticism and judgment are detrimental to the creative process. They are not, they are *instrumental* to the process. Without them your ideas are trivial and frivolous.

So here's how to incorporate what you've learned in this book into the daily practices of your organization. Separate brainstorming into four different meetings, each with a different goal and different set of rules. Brilliantly Borrowed Brainstorming involves: 1) a problem-definition meeting; 2) a borrowing-ideas meeting; 3) a new-idea meeting; and 4) the judgment of these ideas at a separate time.

The first meeting is for defining problems. It's a data dump. Using everything you learned from the first chapter, you work to identify new problems, understand them by finding the root cause, and then define them and the problems that surround them. Looking up and down, you and your team will create the hierarchy of problems that your business works within. This helps you to determine the scope of the problems you want to solve.

Once the first meeting is over, organize your problems by sorting and grouping them. This will help determine if you're missing any problems and it will also provide you with the structure for the next meeting. Once you've got your problem groups, assign different members of your team to different problem groups. Then ask these teams to find other places that have similar problems. These are first competitors, then other industries, and finally places outside the domain of business, like from science or entertainment. Instruct the teams to look for places with a comparable problem.

The second meeting is the borrowing meeting. The teams will get back together and present their research to each another. They'll describe the problem they were assigned and then show how competitors solved it, how other companies solved it, and then how other domains solved it. In essence, you're gathering the materials for the next meeting.

The third meeting is the idea-generating meeting. It's when you begin constructing your solution. Using the material from the last meeting, look for an overarching metaphor for your solution. This metaphor will provide you with a high-level structure for your idea. Try a bunch of different things. Remember, the creative process is a matter of trial and error. This meeting is very similar to the current concept of brainstorming. You may or may not want to suspend the judgment of the ideas, depending on how you think the members of the teams will respond and cultural aspects of your organization (are the egos fragile or not?).

The fourth and subsequent meetings are the evolutionary meetings. These meetings take the most promising ideas and use the three judgment viewpoints to analyze them. This involves debate and the clear delineation of positive and negative attributes. After this meeting, you should have developed one or two promising ideas that need to be assigned to smaller teams for further development. Remember, these things take time; it's a process of trial and error and new problems will arise that require new solutions and so lead to the evolution of your ideas.

At the end of the day, innovation is personal. It happens in the shadows of the mind. These brainstorming sessions are input devices, using members of your organization to help you gather materials, share them, make promising connections, and help to see the positive and negative aspects of possible solutions. However, the inception of an idea will always happen in a deeper and darker place. Don't forget that. It's often easier to come up with a new idea in the early morning, before the sun rises, while your competitors sleep, and your mind is fresh, wiped clean by the night's sleep.

If you're committed to innovation and creativity, then you're committed to the time it takes. You've got to keep trying. To keep climbing. To keep thinking. To keep working on the problem and its solution. Remember, you're a swooper; you're an expert at rewriting, restructuring, and replacing things that don't work. Creative thought is trial and error. There are no magic bullets, just a good grasp of the creative concept and a good grasp of the thinking tools to make it happen. Now it's up to you.

Constructing an idea takes time. Newton developed *Principia Mathematica* over twenty years of intense thinking, identifying problems, gathering data, synthesizing it, and developing solutions to solve his problems. Kepler spent nine years, producing nine thousand pages of notes and calculations, before he concluded that the planet Mars traveled in an elliptical path and not a circular one. Disneyland, from concept to opening day, was an idea that took twenty-five years to construct in Walt's mind. Darwin states in the first paragraph of *On the Origin of Species*—published in 1859—that he began working on it in 1837. Of course, these ideas needed a lot of thought, but if these innovative thinkers had given up, if they had stopped climbing when faced with an obstacle, we would never know their names or the ideas they constructed. There are shortcuts, for sure, but the mountain still needs to be scaled. Failure is not a mistake, it's a waypoint on the way to success. It becomes a mistake if you give up, or as Edison said, "Many of life's failures are people who did not realize how close they were to success when they gave up." You'll never know how close to the summit you were if you give up and turn around. Failure and success are just different stops on the same road.

Of course, you'll often come to a fork in the road. This tends to stop people. Unsure of which way to go, they just sit there and ponder what to do next, instead of choosing a path and giving it a try. Robert Frost gave good advice as to what to do when this happens. He said, "Two roads diverged in a wood, and I— I took the one less traveled by,/And that has made all the difference." You'll always begin your journey on a well-traveled road, borrowing the pathway from others, but eventually the pathway diverges, joins with other paths, and it's at this juncture, if you want to be creative, that you take the road less traveled.

• • • •

The creative process is not a linear one. It's a self-organizing process, and an idea is constructed more in the fashion of a forming thunderstorm than in the step-by-step way an architect designs a building. It's a natural progression that uses itself to construct itself, and so every idea creates its own creative process. Every idea forms a little bit differently. Creative genius does this in the dark shadows of the subconscious mind. Since I lack the natural gift, I begin to realize that I can simulate it even though I don't have it. I realize I can borrow brilliance in the truest sense of the words.

The Sixth Step of the Long, Strange Trip

Tom and I leave Intuit. The opportunity we see lies in online tax software. The IRS doesn't like private companies charging for online tax preparation and electronic filing and so threatens the industry by getting into the business themselves. Scared, the tax prep companies like H&R Block negotiate a compromise. They form the Free File

Alliance, which allows low-income people to prepare and file their taxes for free using private software available at www.irs.gov. The first year millions of people use the site to prepare their taxes for free.

The next few years it grows exponentially. Now four million people are using it. And guess what? It only costs fifty thousand dollars for a private company to join the alliance. That means we can start a company and be listed right along with H&R Block and TurboTax on www.irs.gov, sometimes above them, for the list is rotated so each company has a chance to be on the top. And it's OK to charge the customer for a state tax return, everything's not free, just the federal stuff. We could get hundreds of thousands of paying customers with just a fifty-thousand-dollar investment in marketing. That's an opportunity.

We rent some office space in Oceanside, California, hire a few programmers, construct a Web site, and start building the product. We only have six months to make it. We hire an offshore team in India, and one of the lead developers from Intuit quits and joins us. We call the company TaxNet.

Tom and I fund it out of our personal savings. I've saved about $250,000 and I'm willing to put that into it. Tom's saved a hell of a lot more, but thinks half a million is all we're going to need between the two of us.

"You think we can have the product ready by the first of the year?" I ask Tom.

"Sure," he says. "Piece of cake."

"God, I hope so," I say. I've got everything riding on it. It's like putting all my chips on black and spinning the roulette wheel of life. It feels familiar. And it makes me uncomfortable.

"Have some faith, buddy," he says.

CONCLUSION

THE ROAD LESS TRAVELED

Traveling back in time five years, I see myself in the right front seat of a Cessna 340, we've just taken off, barely clearing the treetops at the end of a dirt runway. I recognize this place as Tamarindo, Costa Rica. Next to me is the pilot, behind me are four friends, half a dozen surfboards, and backpacks full of gear. The pilot speaks broken English and I have a hard time understanding him. Until now.

"Too heavy. Too heavy," he says as he pulls back on the stick, trying to gain altitude, and we head toward the jungled mountains that surround the airstrip. I look back at Johanna. She smiles at me, a smile that contradicts the tears running down her face. We all brace ourselves for the impact.

The belly of the plane scuffs the tops of the trees, we clear the mountains, and so for the time being we're free from danger. We turn west and head toward the ocean. I have a map of Costa Rica on my lap and tell the pilot to head south. We're on a mission. We're in search of the *perfect wave*. Vann, Brian, Chelsea, Johanna, and I have chartered the six-seat Cessna with the intention of flying the west coast, at a low altitude, looking for the ideal surf spot. A hunt for the *perfect wave*. The pilot hasn't filed a flight plan, because we don't know where we're going to land.

We head south, looking for white water, the telltale sign of breaking waves. We pass over Playa Negra. Nothing. Over Playa Hermosa. Flat. Then Jacos, Quepos, and Playa Dominical and still nothing but blue water, no white. Finally, as the minutes tick into an hour, we head out over the Osa Peninsula and the pilot begins to shake his head.

"No Panama," he says in his busted English.

Still no white water. No waves. And now the end of the country is in sight. Disappointed, we decide to put the plane down in the coastal rain forest just north of the Panamanian border on the Osa Peninsula. I point to a small dirt airstrip that is, according to my map, just outside the Corcovado National Park. It looks like there's a small town a mile from the airstrip, on the Gulf of Dulce. Vann and I both know that on the southern tip of this bay is a surf spot called Pavones, which is famous for being, when the swell hits it just right, the longest left-hand wave in the world. It doesn't seem to be breaking today, but we can camp and fish and cross our fingers and hope for a swell.

"What do you think?" I ask Vann.

"Well, if ever there's a chance for the *perfect wave,* I'd say it'd be Pavones," he answers.

So, we land and tell the pilot to pick us up in four days, at the same spot, hoping he understands and doesn't come back in four weeks. I look at Vann and laugh. We're in the jungles of Central America, no place to stay, no sleeping bags, no transportation, and no means of communication. Oh, well. We gather our stuff and begin walking down a dirt road toward the small town I had seen from the air. We hear the deep guttural grunting of howler monkeys in the jungle mocking us as we make our way toward the village.

It's not a friendly place. No one admits to speaking English. It's not a resort town, it's a rural fishing village. There are no hotels. We look suspicious. We feel suspicious. We are suspicious. Finally, someone tells us, in broken English, that there's an American yacht anchored in the bay. Maybe they'd let us sleep on board? Have pity on us. The sun is nearing the horizon, in an hour it'll be dark, and the thought of sleeping in the jungle is beginning to freak me out. I don't trust the howler monkeys and the jungle teems and slithers with life unfamiliar to me. Yeah, a yacht seems like a much better idea.

I look out at the bay, toward the boat, and then back at Vann. I laugh again. You see, there isn't a beautiful American yacht in the water, instead, there's an ancient two-masted ketch, a pirate ship, which looks like it's sailed out of a Steven Spielberg movie. It's not flying a Jolly Roger, but it certainly could be. It looks dangerous, barely seaworthy. However, the sun is setting, behind us the jungle awakens, the soft rustling of things, slithering and sliding, is growing louder and more intimidating and making the pirate ship look like a floating Ritz-Carlton.

We yell out to the boat. The captain hears us and rows ashore. He tells us that he and his wife are sailing around the world, that they left Florida three months ago and are making their way to California. I don't remember his name, but the ketch is called the *Cassiopeia*, after the constellation in the northern sky that sailors have been using to navigate by for centuries. He says we can charter the boat for a few days, he could use the money. So we get into the dory and row out to the pirate ship. We tell him we're in search of the *perfect wave*. He chuckles, says he can help. The sky is a deep red, the color of clotted blood, for the sun has sunk over the horizon and darkness has swallowed the jungle and the Gulf of Dulce.

As we row, something catches my eye, something in the water, as the oar breaks the surface, something amazing, and something hauntingly unnatural. Before I can point it out to the others, I realize they have seen it as well, because Chelsea whispers to us, under her breath, with a sense of awe in her voice, "Oh, my God, did you see that?"

What we see, what's so amazing, is a fluorescent explosion of color coming off of and attaching itself to the oar as it breaks the surface of the bay. For the Gulf of Dulce, on this particular evening, is saturated with phosphorus plankton and every time the oar breaks the water it detonates an eerie green burst of self-illuminating colors. I stick my hand into the water, stir it up, pull it out, and hold it up for the others to see. It glows as if I had just pulled it from a vat of radioactive debris. "Wow," Johanna sighs in disbelief.

That night the five of sleep out on the aft deck of the pirate ship, under a blanket of stars and upon a bed of self-illuminating phosphorus. The next couple of days we sail the bay and the outer reaches of the Pacific, all the time in search of white water, in search of the *perfect wave*. We anchor off of Pavones, praying for a swell, but our prayers go unanswered. It's as calm as a YMCA swimming pool on a Tuesday afternoon. We sail on, like ancient mariners in search of land, looking for the telltale signs of the *perfect wave*. We can't find it. It is not to be.

The fourth morning we begin to make our way back toward the village. Our pilot, if we're lucky, will be waiting for us in the jungle that evening. We stop at the northern point of the bay for lunch and to take in the sun. We sense a small swell starting to boil and see a little white water off the stern. We move toward it. Vann, Brian, and I jump into the sea to catch a few small waves before our rendezvous.

It's a nice little point break, a right-hander. The first wave I catch is amusing, waist high, and I ride it for thirty yards. Far from perfect, but fun. Vann and Brian catch similar ones. The girls relax on the boat, which is anchored just off the point.

Then a set wave rolls through, much bigger than the others. Brian takes off on it, yelling ecstatically as he drops in. Vann and I paddle into position for the next one, but to no avail, it's a single set. We both turn around to see where Brian has ended up, but alarmingly, he's gone. He's disappeared.

"Where is he?" I ask Vann.

"Don't know." He answers.

Holy shit, I think, *Brian's been sucked under.* There's nothing left, both he and his board have vanished. I start to paddle toward shore to look for him when I hear Vann yell at the top of his lungs. "Outside!"

This is a call, familiar to all surfers, for another set wave. This one is bigger and longer than the one that's sucked Brian under. Vann is in perfect position for it and takes off as I struggle to dive under it. I hear him screaming, just as Brian had, as I submerge toward safety.

I come up and look toward the shore. Much to my dismay, Vann has vanished just as Brian did. Now they're both gone. I've been surfing for twenty years and have never seen anything like this before. I've seen people disappear in heavy surf, only to pop up a dozen yards from where they went under. But this is different. The surf is building, for sure, but it's not heavy enough to hold them down for so long. Not even close. Something else has happened to them.

Suddenly, floating in the warm water, I feel strangely naked, vulnerable. I glance back at the pirate ship. Chelsea and Johanna are sunning themselves on the deck, unaware that their shipmates have disappeared below the surface. I sense something in the shadows beneath me.

I look toward the shore and up and down the beach. Where are they? About a mile away I see two people coming toward me. Could it be them? Impossible, I think. How could they be so far away, so far down the beach, when just a moment before they were sitting out here with me? I start to panic. As the two figures come closer I realize, much to my relief, that it is Brian and Vann. They're running, waving frantically, yelling and screaming, looking like two little kids at Disneyland who have just gotten off of Space Mountain for the first time. Questions flash in my mind. What are they yelling about? How the hell did they get so far away? What has just happened?

Neither Brian nor Vann ever explain it to me—it isn't necessary, for a moment later another set comes in, this one for me, I take off, drop in, bottom-turn, then find myself on the shoulder of the most flawlessly shaped, shoulder-high, endlessly long wave I have ever seen in my entire life. I ride it for more than a mile. Speeding down the face, cutting back, and then laughing as it re-forms again, and again, and again.

I have found the perfect wave.

• • • •

Five years later, this story comes to me as I think about thinking. I've taken the creative process out of the shadows of my subconscious mind, made it a conscious process, which, ironically, only makes me a more effective subconscious thinker. Now I understand and it's time to put these thoughts into action, to start applying them to the company Tom and I have founded. But the story about the *perfect wave* gets me thinking: Maybe these ideas are for something different?

Over time, I realize that borrowing brilliance is more than just a way to construct a clever business idea or solve a complex scientific problem, it's a way to solve life's bigger problems as well. As Marie Curie said, "I am my own experiment. I am my own work of art." Or Carl Jung: "But if you have nothing at all to create, then perhaps you create yourself." I realize that my shadow self is telling me the same thing—that these tools can be used to reconstruct myself and the circumstances that define the story of my life.

Living a creative life meant two things to Maxwell Perkins, the legendary editor for Hemingway, Wolfe, and Fitzgerald. He knew people who lived interesting lives, went interesting places, and did interesting things, but who were unchanged by and gained no insight from these experiences. They were not interesting people. On the other hand, he knew people who were very interesting, had deep insights, and seemed to change daily, but never left the comforts of their hometown, never ventured far or did anything remarkable. Led uninteresting lives. Hemingway did both, and Perkins admired him for it. Hemingway traveled to faraway places, did faraway and exotic things, and gained deep insight from them. To Perkins, this epitomized the creative life. To me, it does as well. I long for faraway places and I long to be changed by the journey to these places, whether the journey is physical or intellectual.

Of course, you and I can't live the Hemingway life: big game hunting in Africa, deep-sea fishing in Cuba, and running with the bulls in Pamplona. But you can live a Hemingway-like life. You can borrow Papa Hemingway's attitude—explore, get off the couch and do something important and let the experience change you. Re-create yourself by borrowing the traits from the people you admire. Einstein kept a photograph of James Maxwell on his study wall, alongside pictures of Michael Faraday and Isaac Newton. Bill Gates spent millions to acquire the notebooks of Leonardo da Vinci. And Bruce Springsteen bought his first guitar after he saw Elvis Presley on *The Ed Sullivan Show*. Everyone needs a role model as a source for borrowed characteristics. We construct ourselves just as we construct our ideas just as we construct our lives. And as you know, we can't make something out of nothing, we have to make it out of something else. People are made out of other people. Ideas are made out of other ideas. And lives are made out of other lives—they form the metaphor we borrow to live our own lives.

Borrowing isn't just a creative thinking technique. It's the core technique. Everything creative derives from it. It's why the creative process is so hard to understand. It's why the creative process is so paradoxical. Counterintuitive. The opposite of what you think it is; or what it should be. We are so ingrained with logic, the idea that one thing leads to another, that it's difficult for us to perceive paradox. It's difficult for us to understand things that can't be arranged into neat, point-to-point, linear explanations. The creative process is one of those things. It's full of paradox and it's not very linear.

The Material Paradox

A paradox is something that's self-contradictory. For example—*This sentence is false*—is itself a paradox. If we believe what the sentence says, then we are accepting it as truth, but it says that *the sentence is false,* so it must be false. A paradox. Of course, we could go round and round with this. The sentence defies logic, it contradicts itself. So, too, does the creative process.

Copying is at the core of the creative process. Remember, first you copy, then you create. Copying is how you gather the material to construct your ideas. The only difference between the plagiarist and the creative genius is the source of the material, for they begin in the exact same way. They just steal from different places. Copying and creating, while opposites, derive from each other. Originality results from thievery. This is the material paradox.

Over time, though, the theft gets covered up. It's adjusted, used to solve a different problem and combined with other thefts, and so it gets lost in the DNA of the new creation. The theft is still there, it's just difficult to discern. It's hard to see Lyell's geological theory within Darwin's evolutionary theory. It's hard to see Elvis Presley inside Bruce Springsteen. The evolution of an idea changes the idea over time and the thief's tracks are covered up.

The components you borrow and the places you go to find them determine the creative quality of the solution. Borrow from a competitor and it's larceny. Borrow from outside your field and it's unique and creative. The farther away you venture from the original domain, the more creative your idea will seem. You can also alter the components the way a thief paints a stolen car to hide his theft. For example, you can copy your competitor by doing the opposite to what he does—if he's making a big product, then you can make a small one. You're still borrowing but changing your borrowed material before you use it. Hiding it, so to speak.

And ultimately your idea evolves, changes, through the act of making unique combinations, by putting existing things together that have never been put together before, like the two words *borrow* and *brilliance*. Remember, as screenwriter Wilson Mizner said: If you copy from one, it's plagiarism; if you copy from two, it's research. Combination is the essence of creativity; they are one and the same thing. You can change and alter them, hide your sources as Einstein said, but it's the act of combination that will ultimately drive the evolution of your ideas and that buries the thefts, the copies, deep into the genes of the new things. It conceals the paradox—that in order to create you must copy. Our world is comprised of combinations of earlier combinations. The world is complex and becoming more so with every new idea.

It's exhausting trying to keep up with it all. But ironically, this leads to the second creative paradox. The fountain of youth. The wisdom paradox.

The Wisdom Paradox

Growing old sucks. My back aches. I'm short of breath. And I lose a little more hair on my head each year that mysteriously find its way onto my back. Seems like there's nothing to be gained from growing old. Ah, but there is. There's a paradox to aging. *Let me explain.*

The jaguar has his speed. The elephant his size. The gorilla his strength. And the human being has his mind. We are not the fastest, the biggest, or the strongest, and yet we rule the planet. Unlike any other animal, we can be found in every corner of the globe, not only surviving but reigning over the places and all the species in those spaces, a success undoubtedly due to the magnificent human mind. As Darwin said, "It is not the strongest of the species that survives, nor the most intelligent that survives. It is the one that is the most adaptable to change." We are the most adaptable of all the species and it's the mind that makes us so.

While other species rely on instinct for survival, on their innate wiring system, we have the unique ability to rewire ourselves. Not just to rely on instinct but to override these instincts if we wish. We can reconstruct our minds. Youth is defined by this process. A baby is born in an unwired state, then over the years, thousands of nerve cells begin communicating, then millions, sending messages, creating pathways, then billions, neural networks that form the memories, beliefs, attitudes, and perceptions of the full-grown adult. The only problem is as we get older we get lazy and start using the same networks over and over. This repetition gouges out deep grooves—beliefs set in stone, unwavering opinions, and a longing for the good old days. We grow old. We stop wiring ourselves and just use the same old wires over and over. But we don't have to. Instead we can grow young by continuing the wiring process. By continuing to learn and create. *Let me explain.*

In his book *The Wisdom Paradox,* neurologist Dr. Elkhonon Goldberg says your mind can grow stronger as your brain grows older. He calls this the *wisdom paradox*—that with age comes vigor, not descent—that you can become smarter and wiser by using the networks that living has created to construct new networks instead of relying on them for repetitive thought. In other words, if you keep learning and keep creating you can maintain a youthful mind even as the legs give out and the hair on your head recedes to become the hair on your back. Use it or lose it. To maintain a constant state of rewiring is the essence of youth and is within reach of everyone. As Bob Dylan said, you can become "forever young."

Dr. Goldberg tells of the nuns of the School Sisters of Notre Dame in Mankato, Minnesota. These women are noted for their longevity, living well into their nineties and maintaining vigorous schedules. They read, play games, and take various classes. They have the intellectual stamina of women half their age. What's more remarkable is that none of them appear to have Alzheimer's disease. None show any signs of dementia. Intrigued, doctors decided to autopsy the nuns to see how they were neurologically able to avoid a disease that plagued so many others. Much to their surprise, the doctors found that many of the nuns did have the disease, the tangled plaques and infected brain cells, even though they didn't have the symptoms. According to Dr. Goldberg, this is the result of living a mentally active life, one dedicated to learning, to creating a mind that's in a constant state of construction. When the mind becomes diseased, it simply rewires itself around the infected brain cells and continues to function vigorously. It adapts.

The nuns discovered the fountain of youth. And so can you. It's within the pages of this book. For being creative, like learning, is the ultimate act of rewiring yourself. On his deathbed, the day before he died, Walt Disney described his last idea to his brother, using the ceiling of his hospital room to map it out. Roy Disney would take it and build a newer and better Disneyland in the swamps of Orlando, Florida. Walt died as he had lived, creating till the end, and so living till the end.

Of course, this passion for creativity forms the third creative paradox. The love and apathy of the idea itself.

The Lover's Paradox

"If you love someone," writer Richard Bach said, "then set them free. If they come back they're yours; if they don't, they never were." This is the lover's paradox. Capturing a lover, possessing them, it's said, is not real love. Real love means letting go of the other, letting them fly, not trying to hold on to them or to ground them. The same is true for ideas in the creative process. Capturing them and holding on to them prevents you from letting them evolve. Remember, a kiss is just a kiss. An idea is a one-night stand, not a spouse to move in with, pick out drapes with, and settle down with. You've got to let your ideas go, not become emotionally attached to them, so you can move on to the next one.

This isn't an inherent paradox, it's one to be created. And creating this paradox isn't easy. It's tough love. It's hard to kick your children out of the house. It's hard to throw away something you wanted so badly. It's hard not to love the idea that you constructed, just as it's hard to let your lover go free. People fall in love with the strangest things.

Last year, while skiing, I shared a gondola ride at Mammoth Mountain with a man who spoke with a distinguished southern accent.

"Where are you from?" I asked.

"I live in Atlanta," he said.

"What'd you do in Atlanta?" I asked.

"I'm a medical researcher. I work for the CDC, the Centers for Disease Control," he said.

"Wow," I said. "What kind of diseases do you study?"

He went on to tell me that he specialized in childhood diseases, nasty things like malaria, tuberculosis, and bacterial meningitis. As he spoke I could see the way he admired these things, their intelligence and the clever ways they avoided their own destruction. He spoke of one particularly vicious and deadly strain of meningitis. I could tell how deeply he had thought about it, how well acquainted with it he had become, and the high regard he had formed for it. You see, the deeper you think about something, the more time you spend with it, the more apt you are to admire it, love it, even if it's killing innocent children. Hostages fall in love with their captors—it's called the Stockholm Syndrome, after a famous long-term hostage drama in which the hostages became enamored with their captors. You'll do the same thing with your ideas, even the bad ones. The constant contemplation will create rivers of thought in your mind and you'll get more and more comfortable floating along with them. And once you love an idea, suspended in the river, you won't want to get out. You won't want to change your ideas.

Albert Einstein rejected his own theory of relativity that was used to establish the field of quantum mechanics. Twenty years after he published his famous thoughts, he told a friend, "Yes, I may have started it, but I regard these ideas as temporary. I never thought that others would take them much more seriously than I did . . . A joke should not be repeated too often."

So, you need to let go of your ideas as Einstein did, it's the only way to make room for the next ones, the only way to let them evolve. It's not about the idea, it's about the process. You need to love creating, not the creation. To love the kisser and not the kiss. The journey and not the destination.

And the journey is one big paradox itself.

The Genius Paradox

The creative journey is a matter of constructing an idea by taking it apart, reorganizing it, adding and subtracting things, and then putting it back together. This means creativity requires comfort in chaos, as all of the pieces of an idea scatter in the mind, and a desire for order at the same time. Making a complete mess of things and then carefully, painstakingly, putting it all back together again. The creative thinker is a complete slob at one moment, and then incredibly disciplined and structured at the next. As Nietzsche said, "You need chaos in your soul to give birth to a dancing star."

In preparing his book *Creativity,* Professor Csikszentmihalyi studied hundreds of creative people and saw these paradoxical personality traits. A scientist who was both playful and highly disciplined. A sculptor who began her creation in a state of anarchy and ended the project with incredible control and order. A banker whose office was a mess, desk cluttered with paperwork, and yet he spoke with clarity and developed highly structured financial systems. From chaos comes creative order.

Of course, this is no surprise to us, this is just another case of whole-brained thinking. Using the right brain to perceive things holistically, the overall shape and form; and the left brain to perceive the pieces and how they fit together to form the whole. Taking things apart and putting them back together. The left-brain, right-brain paradox. The genius paradox.

The banger can perceive the pieces while she perceives the whole, making her a genius of creative thought. The swoopers, like us, have to simulate genius, consciously jumping from the left to right and right to left, studying a piece, then looking up to see where it fits into the whole, then putting the head down to see how it connects to another piece. We create stuff through trial and error. We make a new combination, take a step back, see our mistake, take a step forward, fix it, then assess it again until we get it right. The banger does it naturally, in the shadows of the subconscious mind, while the swooper does it consciously. This creates a paradox between the pieces and the whole, between the structure and the components that comprise it.

And a swooper who doesn't know he's a swooper, but thinks he's a banger, creates a paradoxical intention in his mind and so finds it difficult to create anything. Every time he tries to create he draws a blank. I know because this was me.

Let me explain.

Paradoxical Intention

In his book *Man's Search for Meaning*, psychiatrist Viktor Frankl describes paradoxical intention as a kind of reverse psychology in which a person naturally does the opposite of what he intends. A man who tells himself not to sweat and then starts sweating uncontrollably is experiencing it. A woman who tells herself it's impolite to stare and then finds herself uncontrollably staring has it too. My brother and I used to start laughing during Catholic Mass while the priest solemnly spoke of sinners and their fiery destiny. I would tell myself to stop laughing and this would just make it worse, a perverse reverse psychology that Frankl called paradoxical intention. You and I might call it a fight between the conscious mind and the subconscious one. A fight in which the subconscious wins.

Creative thinking tends to bring out this reverse psychology in most people. As soon as you ask yourself for an idea the mind goes blank. Nothing. Your reliance on the subconscious for ideas backfires on you. The shadow self doesn't like to be ordered around. It likes to do its thing on its own time and so punishes you with the sound of silence. Paradoxical intention. Writers call it writer's block and it's deeply feared. It's the severing of the relationship between the conscious and subconscious.

The cure to this disease, in the creative process, is to step in and consciously begin constructing ideas using the thinking tools in this book. Define problems. Borrow from places with a similar problem. Connect and combine the borrowed solutions. Analyze the results and then enhance them by eliminating the weaknesses and exploiting the strengths. What was being done in the shadows can now be brought out into the light. Since you're not a creative genius, you're going to have to go out and find these borrowed ideas. Then consciously construct them. And over time this will teach your subconscious, create mental muscle memory, and slowly any paradoxical intention created by your shadow self will dissipate. It'll eventually join in on the fun.

Like any physical or mental discipline, creative thinking becomes more effective the more you do it. The more ideas you generate, the easier it is to generate the next one. And in the end, as Nobel Prize–winning chemist Linus Pauling said, "The best way to have a good idea is to have lots of ideas."

It's this paradox and the others that add to the layer of fog of misunderstanding that shrouds the creative process. A fog that slowly engulfs even the most effective thinker.

Fogs of Misunderstanding

The creative process is full of misconceptions and paradoxes. Among the most delusional, and most damaging, is the perception that an idea is a stand-alone thing conceived and existing in isolation. It's not. Every idea is part of an evolutionary chain of ideas. Some are direct descendants of others and easy to spot, like the iPod being a descendant of the Walkman. However, some are more divergent, seeming to come from nowhere, unique and original, even though they have descended in the exact same way, like *Star Wars* being a descendant of a Hans Christian Andersen fairy tale. Originality is an illusion. Borrowing materials from a place distant from your subject tends to create this illusion. Likewise, making unique combinations of things tends to cover up the borrowings and the ancestry of the idea. Hiding your sources, Einstein said, is the key to creativity. But this cover-up, making things seem more original than they are, has blurred the true nature of the creative process: its evolutionary nature, one idea being the child of another. In ancient times, this was understood and that is why artists or authors never signed their work. Ideation was thought to be a collaborative effort. It was understood and accepted that ideas evolve from each other, and so there was no reason to claim ownership and no reason for a signature. It wasn't until the free market put a premium price on creative ideas that people started to claim possession of them, giving rise to signatures, trademarks, patents, and copyrights. Originality became a concept born of possession, not a concept of creativity. Today the cult, the illusion, of originality creates a fog of misunderstanding that smothers creativity.

Human nature and social pressures will force you to hide your sources too. You'll become possessive of your own ideas, protecting them the way you protect the PIN number to your debit card. Perhaps you'll even file for trademarks, patents, and copyrights. I've got my name on numerous patents, dozens of trademarks, and a copyright for this book you're reading. This naturally forces my creative process into the darker corners for fear of retribution. If you follow the steps in this book, I promise you, you're going to come up with some good ideas. One of them, if you're lucky, could even bring you fame or fortune. In business, a good idea can be worth millions. I've made a lot of money with some simple and unique combinations of things. But I warn you—you'll begin to forget where the ideas came from, telling people they dropped in from a clear blue sky as you triumphantly file your first patent. Not because you're a thief or trying to deceive, but because that's how they appear to have come to you. This is especially true as you become more proficient at *Borrowing Brilliance* and so relegate much of the process to the subconscious mind. It begins to feel like magic even though it's no different than learning how to drive a car. I tell you this only to remind you that at the heart of the creative process is borrowing. Remember: Your ideas are the children of other ideas. If you keep this in mind, then the fogs of misunderstanding will dissipate.

• • • •

Now, as you venture out on your own creative journey, you know you're in search of an idea. Like the search for the perfect wave, this is an adventure. If you keep looking, take the road less traveled, you may find yourself on a pirate ship, in a phosphorous bay, with the longest wave in the world as your playground. Don't just sit there and wait for it to come to you. You need to go to places with a similar problem and look for ideas to borrow. That's how every idea is constructed, whether it's a business idea, a scientific idea, an entertainment idea, or an idea for a new casserole recipe. There are no truly original thoughts. Originality lies in the construction of other conceptions, for brilliance is borrowed.

Always is. Always was. And always will be.

Go figure, right?

The Final Step on the Long, Strange Trip

The six months till tax time click by quickly. We hire Stacia, John, and Fred to convert the tax code into logic for the programmers in India. Under Tom's direction, the lead developer from Intuit creates a software architecture, an underlying structure for the product. Kim and I lay out the Web site and the user interface for the product. Chris designs it.

A month before launch we run out of money. I've invested all of my savings. This is a bigger project than we thought. I take a second mortgage out on my house, put it all on black, and let the roulette wheel of life spin again. We keep plugging away.

A couple weeks before launch the lead developer quits. He sends us an e-mail, and tells us it's never going to work. He says we took on too much, too fast, and don't have the resources to make it. I want to form a lynching party and go over to his house and hang his scrawny ass from the palm tree in his front yard. Tom calms me down, saying he doesn't think that's a good idea.

"Don't worry," Tom says, "have faith, we're going to make it."

I start drinking again in the evenings. This seems so familiar to me. I feel the quicksand at my feet; feel myself being sucked into it again. Mohan, a programmer we hired in California from the off-shore team, steps in and replaces the lead developer. He echoes Tom's sentiments. "Don't worry, have faith, we're going to make it."

Launch day, January 1, rolls around. We run the program. I test it myself. There are 348 errors in the system. *Holy shit!!!* We can't launch with all these errors. I send the error list to India. They fix the errors. A couple days later we run it again. Now there are 612 new errors. *We're going in the wrong direction.*

We're running out of time and we've run out of money again! Tom puts his half in for the next round, but since I've used up all my savings and mortgaged my home I have to call my ex-wife—Terri—and tell her I need some money. I tell her it's just a short-term loan. That we're going to make millions once we launch the software.

"Are you sure about this?" she asks.

"Don't worry," I echo, "have faith, we're going to make it." Incredibly, but in line with her character, she uses a line of credit secured by a second mortgage on her house to wire me the cash. *Go figure, right?* I put her money on black, with mine, with Tom's, and we spin the wheel again.

A few more weeks go by. We know that the tax season peaks on the last day of January. We make that our new goal. We all pull together. Stay up all night, work all day, and do it again. Finally, we try it once more. I test it myself. No errors!!!!

"Let's launch it!" I scream.

"Okay," Tom says. "But remember, we never know what's going to happen until we get it under a live load, with hundreds of people using it at the same time."

"I hope that's the problem," I say.

Tom and Mohan prepare the program, do the voodoo they do. Connect it to the IRS site. And we go live. Everyone stands behind Mohan as they watch what happens. I can't bear to look. Everything's riding on black: my savings, my home, my ex-wife's home, Tom's savings, everything. I sit in the corner and look at the faces as they watch the screen. Mohan clicks away on the computer. I have no idea what he's doing. Tom is sitting next to him. Faces are blank. Poker faces. It's quiet for a moment. Then.

"Holy shit," Tom says. "Hey, buddy, you gotta come here and see this."

WHAT A LONG, STRANGE TRIP

Lately it occurs to me
What a long, strange trip it's been.
—JERRY GARCIA

As I crest Kingsbury Grade, Lake Tahoe unfolds below me. Bigger, brighter, and bluer than I have ever seen it. I pull over and get out of the car. It's been five years since I've stepped into the blue world and I have missed it dearly, as much as anyone can ever miss a place. As I stand here I realize that I've never seen the lake from this particular angle. There are half a dozen roads that lead into the basin, but I've never entered by Kingsbury Grade before. It takes my breath away. It's more beautiful than I even imagined.

It's now six months after the launch of our online tax software, and a few days earlier I had signed the papers that successfully sold TaxNet to the tax preparation giant H&R Block. What Tom had wanted me to see when he and Mohan took the site live was that hundreds of users were logging on in the first few moments. Although we missed a huge part of the tax season, we still managed to complete and electronically file over 130,000 tax returns for the season. This drew the attention of H&R Block and they offered us millions of dollars for the program and for us to come and run the tax software division of their company. It wasn't the fifty million that I had left on the table five years earlier, but it was enough to pay off some debts, buy a beautiful cabin in Tahoe City, and so complete the journey, the long strange trip back home. This time the roulette wheel of life had landed on black. This time I had won.

That afternoon I return to Jake's on the Lake. I take my usual seat at end of the bar, in the corner, and near the exit, even though I don't anticipate needing to make an easy escape. I have returned triumphantly, to celebrate with my friends, Stoli and Cran. Tim, the bartender from five years before, turns and looks at me. He doesn't seem surprised to see me.

"The usual, Dave?" he asks.

"That's right, Tim. And keep 'em coming," I answer.

At first I think he's joking. It's been five years since I'd been in the place—yet he's acting like I was in here yesterday. He hands me the drink, smiles, and turns to serve another customer. He's not joking. He doesn't know that I've been gone for five years. So much water has washed under the bridge and I've been swept away with it, but somehow I've managed to float right back into the same spot. Weird. He has no idea how happy I am to be back here.

Amused, I pick up my drink and walk out onto the deck behind the restaurant. It's a crisp spring day and there isn't a cloud in the sky. As I stand and admire the lake and the peaks that surround the water, it suddenly occurs to me that I've never climbed any of them. I count. There are eighteen different summits, several over ten thousand feet, and a dozen over nine thousand. I gaze out at Freel Peak at the south end of the lake, the highest point in Tahoe, and I wonder what it would be like to be up there looking down.

I walk back into the bar. Put my Stoli and Cran on the counter and lay a twenty-dollar bill next to it and walk out. It's not until years later, when I sit down to write these words, that I realize I never even took a sip to celebrate my return. I simply forgot to.

Go figure, right?

David Kord Murray
Tahoe City, California

APPENDIX A

SUMMARY OF THE SIX STEPS

A Two-Page Summary of the Six Steps to Constructing a Creative Idea

SUMMARY OF THE SIX STEPS

Step One: *Defining—*

> *Define the problem you're trying to solve.* A creative idea is the solution to a problem. How you define it will determine how you solve it. Mistakes result from solving too narrow or too broad a problem. So, identify as many problems as possible using tools like observation and then sort from high-level to low-level problems.

Step Two: *Borrowing—*

> *Borrow ideas from places with a similar problem.* These are the construction materials for your solution. Using your problem definition, borrow from places with a similar one, so start with your competitors, then look to another industry, and finally look outside business and to the sciences, arts, or entertainment to see how they solve that problem.

Step Three: *Combining—*

> *Connect and combine these borrowed ideas.* Making combinations is the essence of creativity. So, using the borrowed materials from the last step, find an appropriate metaphor to structure your new idea. In other words, use an existing idea to form the framework for a new idea by establishing a metaphor, extending it, and then discarding it when it no longer works.

Step Four: *Incubating—*

> *Allow the combinations to incubate into a solution.* The subconscious mind is better at making combinations. To do this, give the subconscious time to work and quiet conscious thought so you can listen to the subconscious speak. Use tools

like: sleeping on it, pausing, putting it away, and listening for misunderstandings. In other words, often the most effective thinking is not thinking at all.

Step Five: *Judging—*

Identify the strength and weakness of the solution. Judgment is the result of viewpoint. Intuition the result of judgment. Use positive and negative judgment to analyze your solution and identify the strengths and weaknesses of the idea. This leads to creative intuition: an idea that has these things (positives) but not those things (negatives).

Step Six: *Enhancing—*

Eliminate the weak points while enhancing the strong ones. Ideas evolve through trial-and-error adjustments. They self-organize. Return to the first five steps to make your adjustments of the idea: redefine; reborrow; recombine; reincubate; and rejudge it all. The order in which you do these things will be different with every idea, for the creative process will create itself.

APPENDIX B

SUGGESTED READING LIST

FOR FURTHER STUDY ON THE SUBJECTS CONTAINED IN THIS BOOK

Suggested Reading List

Hundreds of books went into the writing of *Borrowing Brilliance*. This is not a complete bibliography, these are just my favorites. Start here if you'd like to learn more about the concepts of creativity contained in *Borrowing Brilliance*.

Things That Make Us Smart: Defending Human Attributes in the Age of the Machine by Don Norman and Tamara Dunaeff (New York: Perseus Books, 1993). Don explains that we use artifacts to make us smarter, physical things like a computer or mental things like the multiplication tables. My thinking tools are homage to this concept so eloquently explained here. Don is a great teacher and a great writer.

The Thinker's Toolkit: 14 Powerful Techniques for Problem Solving by Morgan D. Jones (New York: Random House, 1995). This is the best book available if you want to learn more about the problem itself. Understanding it and defining it. Chapter One in *Borrowing Brilliance* is written in deference to this one. The first half of this book is the best and most applicable to our subject.

A Technique for Producing Ideas by James Webb Young (New York: McGraw-Hill, 2003). This "pamphlet" was written in the 1940s by advertising executive James Young. I discovered it after I had written the first draft of *Borrowing Brilliance* and was surprised how close it came to my process. It's short; you can read it in less than an hour. A nice introduction to creative thinking.

Six Thinking Hats by Edward de Bono (Boston: Little, Brown and Company, 1985). This book is the most often quoted creativity book ever written. The premise is that you need to adopt different personas, different hats, for the different parts of the creative process. Simple to read, short, and a must for anyone studying creativity.

A Whack on the Side of the Head by Roger von Oech, Nolan Bushnell, and George Willett (Menlo Park, Calif.: Creative Think, 1973). This is a fun book and yet insightful at the same time. It's full of different thinking techniques, things you can do to come up with new ideas. Lots of good examples too. A must for the creativity student.

On Writing by Stephen King (New York: Scribner, 2002). Stephen is a banger, a genius, and this book is a fun read, a tour of a demented mind and the craft that he's perfected. He explains the source of his ideas even though he says, at the beginning of the book, that he doesn't know where his ideas come from. This has creative applications for any subject, not just writing. And fun—King can write, of course.

Flow: The Psychology of Optimal Experience by Mihaly Csikszentmihalyi (New York: Harper & Row, 1990). This is an analytical/psychological explanation to getting "into the zone." Being exceptionally creative requires being in the zone, the place where conscious and subconscious thought meld into one. Interesting read.

Creativity: Flow and the Psychology of Discovery and Invention by Mihaly Csikszentmihalyi (New York: HarperCollins, 1996). This is the seminal work on creative thinking and the creative thinkers. Mihaly did an extensive study of hundreds of creative people from art, science, and business and this book is the result of that study. Long, academic, but gives you great insight into the creative process and the creative thinkers. I borrowed a lot from this book. Good writing style too.

Changing Minds: The Art and Science of Changing Our Own and Other People's Minds (Leadership for the Common Good) by Howard Gardner (Boston: Harvard Business School Press, 2006). Being trapped in repetitive thought is the curse of the uncreative mind. In this book Gardner describes the mind and how you can alter, modify, and transform your thought patterns. Your mind is your thinking tool, this helps you to understand it.

How the Mind Works by Steven Pinker (New York: W. W. Norton Company, 1999). It's hard to put this book down. Pinker uses practical metaphors to explain the complex workings of the mind. These are things you need to know if you want to become more adept at using it. Your mind is the architect of the creative idea. Long but entertaining. Pinker's cool.

Synaptic Self: How Our Brains Become Who We Are by Joseph LeDoux (New York: Viking, 2002). A good introduction to neurology, but a little bit hard to follow at times, deep. He explains how memory is a pattern of nerve cells connected by the billions of synapses. This wouldn't be the first "brain" book I'd read, but should be on the list once you've got a basic understanding of neurology. Well written.

Emotional Intelligence: Why It Can Matter More than IQ by Daniel Goleman (New York: Bantam Books, 1995). Understanding and using emotions are an important part of the creative process. This book does a great job of explaining them. Easy to read and filled with practical examples.

Man's Search for Meaning by Viktor E. Frankl (Boston: Beacon Press, 1959). Every creative thinker needs to be passionate. Frankl explains that finding meaning in something, anything, is the key to psychological health. He discovered this as a concentration camp survivor and details those experiences. Compelling read, you can't put it down.

Basic Freud: Psychoanalytic Thought for the 21st Century by Michael Kahn (New York: Basic Books, 2002). This is an easy-to-read summary of the concepts put forth by Freud. It gives a great explanation of Freud's banquet hall/drawing-room metaphor for subconscious/conscious thought. A good introduction to psychology. Fun read.

The Power of Now: A Guide to Spiritual Enlightenment by Eckhart Tolle (Novato, Calif.: New World Library, 1999). Quieting the conscious mind is critical to creative thinking. Not thinking is effective thinking. This book explains that being in the present moment is a means to quieting the mind. Easy to read and understand.

Doubt: A History: The Great Doubters and Their Legacy of Innovation from Socrates and Jesus to Thomas Jefferson and Emily Dickinson by Jennifer Michael Hecht (New York: HarperCollins, 2003). I usually don't like long, academic-type books, but this is an exception. Jennifer is a poet and so the prose is magnificent and it gives the best history of innovation I've ever read. Makes you proud to be human. Long but worth it.

The Reluctant Mr. Darwin: An Intimate Portrait of Charles Darwin and the Making of His Theory of Evolution (Great Discoveries) by David Quammen (New York: W. W. Norton, 2006). This is the best introduction to Darwin and Darwin's ideas. Well organized. Well thought out. You'll get a great overview of evolutionary biology without getting lost in the details. And it's a great story of a great life. Read this before you read the two books that follow below.

The Selfish Gene: 30th Anniversary Edition by Richard Dawkins (Oxford, New York: Oxford University Press, 1989.) A must-read for anyone who loves nonfiction. This is as much a work of literature as it is a work of science. It is the gene, Dawkins proposes, that fights for survival and not the organisms that contain it. We are replicating machines for the gene, we ensure its survival. This book is brilliant. A must read.

On the Origin of Species: By Means of Natural Selection (Dover Thrift Editions) by Charles Darwin (Mineola, N.Y.: Dover Publications, 2006). Written over a hundred years ago, it's still a compelling read. Perhaps the most important book ever written. This is an abstract; Darwin planned on publishing a more complete work on the subject, but it turned out so good that he never wrote the complete work. Even if you don't believe, read it for literature; however, it's hard not to believe his theory after you read this book.

Isaac Newton: The Last Sorcerer (Helix Books) by Michael White (New York: Basic Books, 1999). Newton is one of the most creative minds that have ever lived. This book does a good job of telling the story of his life and how he developed his theories. Newton spent a good part of his life studying the Bible and alchemy, so I mostly skipped those parts of the book. Makes you wonder what he could have done if he had applied himself more to the practical sciences.

Einstein: His Life and Universe by Walter Isaacson (New York: Simon & Schuster, 2008). Einstein was pure genius. He was also an interesting character. While long, this is a well-written book and tells the stories behind the man.

iCon Steve Jobs: The Greatest Second Act in the History of Business by Jeffrey S. Young and William L. Simon (Hoboken, N.J.: Wiley, 2005). Jobs is a creative genius. He's not a one-trick pony, a guy with just one great idea— he understands the creative process. This is a good history of Jobs and how he became the most successful businessman of our era. You can't be a student of creativity without being a student of Steve.

The Innovator's Dilemma: The Revolutionary Book That Will Change the Way You Do Business by Clayton Christenson (New York: Collins,1997). This book is about business innovation and claims that to innovate, a business must destroy its current business—thus creating the "dilemma." A good book to read if you work for a large company.

Marketing Warfare: How to Use Military Principles to Develop Marketing Strategies by Al Ries and Jack Trout (Advanced Management Reports, 1986). One of the best books on business strategy, written by the guys who coined the term *positioning*. Easy to read and understand but gives you deep insight into business, marketing, and strategy. A must-read for the business innovator, you can't come up with business ideas unless you understand business.

The Hero with a Thousand Faces (Bollingen Series) by Joseph Campbell (Novato, Calif.: New World Library, 1949). This is the book that George

Lucas used to construct the *Star Wars* screenplay. While sometimes hard to follow, it's the best example of borrowing and how it's inherent in the way we think. Worth the struggle. If you really want to read the best of Campbell, though, download his lectures or the Bill Moyers interview from iTunes.

Metaphors We Live By by George Lakoff and Mark Johnson (Chicago: University of Chicago Press, 1980). Creative thinkers are metaphorical thinkers. It's the metaphor that we use to construct new ideas and this book describes how metaphors naturally structure our thoughts. It's a classic. A must-read.

ACKNOWLEDGMENTS

It's difficult changing careers, especially later in life; we tend to get comfortable doing what we've always done. One thing I've learned, over the years, is that you need to surround yourself with great people if you want to succeed. That's what I've done as I try to switch careers, try to become a writer. I'd like to thank Don Norman, one of my favorite authors, for his advice and for introducing me to Sandy Dijkstra, his agent, who took a chance on me and became my agent. Sandy and her team, Elise Capron and Jill Marsal, not only sold this book but helped to focus my thoughts and form the overall shape of the manuscript. Sandy introduced me to Larry Rothstein, who helped me to write the proposal for this book and then stayed with me for the entire writing process, offering encouragement when I was down, insight when mine was off, and keeping me structured while allowing me to create at the same time. Larry was my guardian as I wandered through the darkness of writer's hell and finally back to the light with a completed manuscript. Then there's Bill Shinker, my publisher at Gotham Books, who taught me how to unpack a manuscript— making my ideas easier to understand and the book more enjoyable to read. Finally, thanks to Jessica Sindler, my editor at Gotham, who took the time and trouble to carefully explain the problems with my first draft and guide me through the painful process of rewriting the second and third ones (which were really the fifth and sixth drafts, but don't tell her). Jess is great.

This book is a thinly disguised memoir, the story of my life, and what I've learned as I've tried to live a creative one. In that story three people changed my business career, setting me on a new course, a new direction in life, more than any others. The first was John Murray, my dad, who talked me into getting a degree in structural engineering, forever changing the way that I think and giving me the confidence to think boldly. The second was Chris Broom, my brother-in-law, who gave me the capital to start my first company, forever changing my career direction and giving me the confidence to do bold things. The third was Tom Allanson, who rescued me after I lost my company and life savings and gave me a new start when I needed one more than he will ever know. These three men changed my life for the better and I am grateful.

Then there are the people I admire. The people whom I've borrowed from to create myself. Many of the ideas in this book are a direct result of having known them, listened to them, and become them. First, there's Nancy Murray, my mom, whom this book is dedicated to and who taught me to get off the couch and do something, anything, to go out and live life. I wish I could be what she thinks that I am. Then there's Katie Murray, my daughter, who teaches me every day about living and having fun and establishing friendships. There's Terri Murray, Katie's mom and my ex-wife, who's always supportive, even though we're no longer married. Louis Schneider, my friend and business partner at Preferred Capital, whose sense of humor and general outlook on life still make me laugh even though I haven't seen him in years. Randy Brecher, my Tahoe friend, who has more focus than anyone I know and who tries to pretend he's a grown-up but doesn't know that I know he isn't. Camille Miller, my friend, who likes to be different just for the sake of being different, and I like that about her. Deborah Prince, who helped me to work out some of the stories in this book and taught me to get up early and make something out of the day because that's what she did. And finally there are my two best friends, who together with me form *The Three Amigos*: David Meyers and Kimberly Benintendo. Dave inspires me with his love for life and the energy he brings to living. Kim inspires me with the passionate and unconditional way she leads her life; she has, with the exception of family, stood by me more than anyone else, through the heady highs and the lowly lows, never wavering in her stance. No one could ever wish for better friends. The best parts of me are borrowed and found in the best parts of my family and friends. *Thank God.* I am deeply in debt to all of them.